T0336141

Tangles

Tangles offer a precise way to identify structure in imprecise data. By grouping qualities that often occur together, they not only reveal clusters of things but also types of their qualities: types of political views, of texts, of health conditions, or of proteins. Tangles offer a new, structural, approach to artificial intelligence that can help us understand, classify, and predict complex phenomena.

This has become possible by the recent axiomatization of the mathematical theory of tangles, which has made it applicable far beyond its origin in graph theory: from clustering in data science and machine learning to predicting customer behaviour in economics; from DNA sequencing and drug development to text and image analysis.

Such applications are explored here for the first time. Assuming only basic undergraduate mathematics, the theory of tangles and its potential implications are made accessible to scientists, computer scientists, and social scientists.

REINHARD DIESTEL is Chair of Discrete Mathematics at Universität Hamburg, where he works on structural graph theory and combinatorics. He is the author of the leading graduate-level text *Graph Theory* (6th edition, 2024), which has been translated into German, Russian, Japanese, and Chinese.

Tangles

A Structural Approach to Artificial Intelligence in the Empirical Sciences

Reinhard Diestel

Universität Hamburg

Shaftesbury Road, Cambridge CB2 8EA, United Kingdom

One Liberty Plaza, 20th Floor, New York, NY 10006, USA

477 Williamstown Road, Port Melbourne, VIC 3207, Australia

314–321, 3rd Floor, Plot 3, Splendor Forum, Jasola District Centre, New Delhi – 110025, India

103 Penang Road, #05–06/07, Visioncrest Commercial, Singapore 238467

Cambridge University Press is part of Cambridge University Press & Assessment, a department of the University of Cambridge.

We share the University's mission to contribute to society through the pursuit of education, learning and research at the highest international levels of excellence.

www.cambridge.org
Information on this title: www.cambridge.org/9781009473316

DOI: 10.1017/9781009473323

First published 2024

A catalogue record for this publication is available from the British Library

A Cataloging-in-Publication data record for this book is available from the Library of Congress

ISBN 978-1-009-47331-6 Hardback

Contents

Part II. Tangles in Different Contexts
A collection of informal examples

Part III. The Mathematics of Tangles
Concepts, theorems, algorithms

Part IV. Applying Tangles
Back to the examples

Preface

This is a book about potential applications of a new mathematical theory, written by a mathematician for a non-mathematical readership. Its style develops from an intuitively informal to a more formal level that uses basic mathematical language, just enough to make things precise. No serious mathematics is used anywhere in the main body of this book.[1]

This preface says a little about where tangles come from in mathematics, so as to indicate what is new in this book and what is not. Readers without this background are encouraged either to just skim the preface for a quick impression, or to skip straight ahead to Chapter 1. This begins with three separate introductions addressing natural scientists, social scientists, and computer scientists in turn.

The mathematical theory of tangles has its origins in the theory of graph minors developed by Neil Robertson and Paul Seymour in the final two decades of the 20th century. In a series of over twenty research papers, which culminated in the proof of one of the deepest theorems in graph theory, the *graph minor theorem*, Robertson and Seymour developed a new connectivity theory tailored specifically to the somewhat 'fuzzy' notion of their central object of study, that of a graph minor. Their new connectivity theory centred around a revolutionary new concept of high local connectivity in a graph: the notion of a *tangle*.

Loosely speaking, a tangle is a region of a graph that hangs together in an intricate way. Intricate in that, while being close-knit in the sense of being difficult to separate, it does not conform to the usual graph-theoretic notions of high connectivity.

Tangles constituted a shift of paradigm in what high *local* connectivity, somewhere in a graph or network, might mean. There is a standard measure of global connectivity for graphs, and the traditional way to measure their local connectivity was simply to look for regions in the graph that were highly connected in this global sense, applied to the region as a subgraph. As these highly connected regions were themselves viewed as graphs, they would be decribed in the same way as graphs are: by precisely naming their 'vertices' (or 'nodes') and the 'edges' connecting them.

Graph minors, on the other hand, the objects that Robertson and Seymour set out to study, are fuzzier substructures than subgraphs: a highly connected minor will usually persist even if the graph containing it is changed a little. Rather than describing these minors in the traditional, somewhat pedestrian, way of naming all their vertices and edges, Robertson and Seymour thought of an ingenious indirect way to capture just their essence: by declaring for every bottleneck in the graph on which of its two sides *most* of that minor lies.[2]

Such a collection of pointers at the bottlenecks of a graph came to be called a *tangle*. Of course, this is a hugely abstract kind of thing – if indeed it merits being called a 'thing' at all. However, bundling even the most complicated collections of objects and their relationships into a single notion is a process not uncommon in mathematics: it enables us to move on and describe more concisely any higher-level structures in which such composite objects occur. In our example, the collection of pointers that constitute a tangle in a graph, one at each of its bottlenecks, deliberately ignores the detail of what vertices and edges our highly connected minor consists of. Instead, it just records where most of it lies – relative to *every* bottleneck.

It turned out that this deliberate restriction of information about the highly connected minors in a graph came with a gain in clarity: the detail discarded was clutter, the information retained its essence.[3]

This development in graph theory was followed by a discovery which, quite unexpectedly, made the entire theory of graph tangles available for the analysis of highly cohesive substructure far beyond graph theory: it turned out that, not just for the notion of a tangle but also for the proofs of the deepest theorems about them, it is enough to know the relative position of those bottlenecks, rather than how exactly they divide the graph of which they are bottlenecks. This information can be encoded in some abstract way that is quite independent of graphs.

The theory of tangles has thus become applicable to a wealth
of real-world scenarios. The purpose of this book is to show
how this can work.

Our narrative starts with a naive discussion of what tangles mean in various real-world scenarios, and how tangle theory can make an impact there. It then takes the reader through the basic mathematical underpinnings of abstract tangle theory, just enough to enable them to set up a rigorous quantitative framework for applying tangles to their own field. It winds up by revisiting the example scenarios to show how the more formal theory plays out in these contexts.

It should be stressed that those real-world scenarios discussed are highly simplified: they are toy examples of contexts in which tangles can be applied. In reality, they can probably be applied somewhere in most of the natural and quantitative social sciences. This will require the input of experts in those fields. It is the aim of this book to put such experts in a position to try this out for themselves; generic software for this purpose is available via `tangles-book.com`.

The layout of this book is as follows. It begins in Chapter 1 with three short introductions to what tangles are, and what they are designed to achieve: in the natural sciences, in the social sciences, and more specifically in data science. These introductions can be read independently of each other, and are written so as to appeal to readers with these respective backgrounds. In this way, they provide three separate entry points to this book. However, they show aspects of the same big picture, and none of them requires any expertise in the area for which it was written. Hence readers with any background may well benefit from reading all three of them: they are all short, and they illuminate the notion of tangles from three rather different angles.

Chapter 2 develops the notion of a tangle from the intuitive picture formed in Chapter 1, still at an informal level. This will be accomplished by the end of Section 2.3. At this point, any reader who cannot wait to see some tangle applications will be sufficiently equipped to skip ahead to Part II, where applications are discussed informally on the basis of just the notion of tangles, not their theory as described later in the book.

Chapter 3 gives a first indication of the two main theorems about tangles, still not in formal mathematical language but in terms of the example settings described in the three introductions. Together, Chapters 1–3 form Part I of the book, an informal introduction to the notion and theory of tangles from three rather different application perspectives.

Part II continues with a collection of explicit example scenarios in which tangles might be applied, and describes informally what the mere notion of a tangle can already achieve there. As pointed out earlier, these example scenarios are highly simplified and, in their simplicity, artificial. The idea behind going through such a range of examples is to indicate the potential of tangles throughout the sciences, and to do so in an unassuming way that inspires readers to find tangles in their own field of expertise.

The examples in Part II were chosen to illustrate the diversity of potential tangle applications. The corresponding chapter sections may be dipped into at liberty: nothing here is required reading for any material later in the book, except for the corresponding sections in Part IV.

Part III then explains tangles a little more formally. It still does not assume any knowledge of advanced mathematics, but the description is in basic mathematical terms such as sets, subsets, functions and so on. The idea is that this more formal description of the notion of tangles, given in Chapter 7, should be precise enough to enable the reader to apply tangles to their own individual background.

Chapter 8 continues with statements of the two main tangle theorems. The first of these describes how the tangles of a large dataset lie with respect to each other: how some tangles refine others, and how the most refined tangles are separated by some particularly crucial bottlenecks which, between them, organize the dataset into a tree-like shape that displays where its main tangles lie. The second fundamental tangle theorem, which is equally important, tells us what our data looks like if it has no tangles. It offers verifiable quantitative evidence of the *lack* of structure in our data – for example, if it is polluted or inconclusive for some other reason.

The remainder of Part III describes the mathematical toolkit that enables us to tune tangles to fit an intended application (Chapters 9–10), and then describes the fundamental tangle algorithms in Chapter 11.

In Part IV, finally, we return to the examples discussed informally in Part II. Equipped with the formal notions from Part III, and having met the two main tangle theorems, readers will be able to see not just what tangles mean in those various contexts, but also how they can be structured and fine-tuned to offer insights relevant to that field.

Throughout the text, there are markers for 'footnotes' that are collected together at the end of the book. The reason I have implemented these as endnotes is that they can happily be skipped at first reading: they offer further illustrations, more detailed explanations and so on, which are not meant to interrupt the flow of reading unless the reader feels curious for more at that point already.

This book would not exist but for the inspiration and contributions in substance I received from numerous people over the past few years. The development of abstract tangle theory that underlies the applications envisaged here began with an idea of my student Fabian Hundertmark, who extracted from our then recent proof [6] of a canonical tree of tangles theorem for graphs the algebraic core of tangles that was actually needed in that proof [26]. When Sang-il Oum visited me in 2013, we found a proof also of the tangle–tree duality theorem based just on these minimalist algebraic foundations for the notion of a tangle [14, 15]. This set the scene for the development of abstract tangle theory based on [9], which was carried through in the following years mostly by various members of my Hamburg group, particularly by Sandra Albrechtsen, Johannes Carmesin, Christian Elbracht, Ann-Kathrin Elm, Raphael Jacobs, Paul Knappe, Jakob Kneip, Max Teegen, Hanno von Bergen and Daniel Weißauer.

The idea to use this abstract theory of tangles for applications outside mathematics was born when I told Geoff Whittle about it in Oberwolfach in 2016. I remember vividly his exclamation, '*surely*, as we can see structure and things in images so quickly, our brain just sees tangles!'. We then proved that in [16], most of which is now part of Section 14.6.

In the years that followed I benefited immensely from discussing tangle applications with quite a diverse set of people. Outside mathematics these include Partha Dasgupta in economics, Jane Heal in philosophy, Thomas Günther in virology, Chin Li in psychology, the CNRS group around Oliver Poch and Julie Thompson in protein sequencing, Rolf von Lüde in sociology, Ulrike von Luxburg in machine learning, as well as the people at Google including, in particular, Krzysztof Choromanski. Within mathematics they include Nathan Bowler, Joshua Erde, Jim Geelen, Rudi Pendavingh, and Geoff Whittle. To all these I extend my thanks for their ideas, enthusiasm and encouragement!

Last but not least, I thank my tangle software group of Dominik Blankenhagen, Michael Hermann, Fabian Hundertmark and Hanno von Bergen for their amazing success in bringing this pie down from the sky and rooting it firmly in fertile earth. Their generic tangle software is now available via `tangles-book.com` under an open-access licence [1]: for all who would like to play with it or just see some examples in action, to apply it in their own professional context, or to develop it further by adding their own packages to the library.

Reinhard Diestel, February 2024

Part I

Tangles

A new paradigm for clusters and types

This first of the four main parts of the book offers a gentle and informal introduction to tangles.

We set out, however, not from what tangles are, but from where they might take us: what difference they might make to some fundamental methodological approaches in the natural sciences, in the social sciences, and in data science. Chapter 1 offers three separate introductions for readers from these three backgrounds. They can be read independently, and in any order. Readers are encouraged to read them all. For not only do they reflect the diversity of potential tangle applications, but through this diversity also highlight seemingly unrelated aspects of the notion of a tangle, the concept central to this book.

The notion of a tangle is then developed informally in Chapter 2. There will be ample reference back to the introductions from Chapter 1, to relate the rather abstract concept of tangles, as it is slowly developed, to their potential applications right away. Readers keen to see some concrete examples of how tangles might be applied, but less curious about the various aspects of the notion as such, may skip ahead to Part II after Section 2.3.

Chapter 3 offers a glimpse of what tangle *theory* has to offer in addition to just the notion of a tangle. The latter, however, goes a long way towards many tangle applications already. Chapter 3 can therefore be skipped by readers who feel sufficiently dedicated to read about tangle theory and its uses in Part III, where both the notion and the theory of tangles are developed rigorously at their simplest mathematical level.

1

The idea behind tangles

This chapter offers three introductions to the concept and purpose of tangles: one for the natural sciences, one for the social sciences, and one for data science. These introductions can be read independently, and readers may choose any one of them as an entry point to this book, according to their own background.

However as all three introductions illuminate the same concept, readers from any background are likely also to benefit from the other two viewpoints. Indeed, while each of them may seem plausible enough on its own, they are rather different. The fact that they nevertheless describe the same concept, that of a tangle, illustrates better than any abstract discussion the breadth of this concept and its potential applications, including in fields not even touched upon here. Moreover, even in a given context where one of the three viewpoints seems more fitting than the other two, switching to one of those deliberately for a moment is likely to add insight that would otherwise be easy to miss.

1.1 Tangles in the natural sciences

Suppose we are trying to establish a possible common cause of some set of similar phenomena. To facilitate this, we may design a series of measurements to test various different aspects of each of these phenomena.[1]

If we already have an overview of all the potential causes, we might try to design these measurements so that each potential cause would yield a list of expected readings, one for each of these different measurements, so that different potential causes differ in at least one measurement.

Then only the true cause would be compatible with all the readings we get from our actual measurements performed on the phenomena we are trying to explain.

In our less-than-ideal world, it may not quite work like this. For a start, we might simply not be aware of all the potential causes – not to mention the fundamental issue of what, if anything, is a 'cause'. Similar phenomena may have different causes, or no single cause. Even potentially single causes need not be mutually exclusive but may be able to co-exist; then we shall not be able to design experiments that will exclude all but one of them with certainty. And finally, measurements may be corrupted, but we may not know which ones were.

We usually try to compensate for this by building in some redundancy: perhaps by taking more measurements, or by measuring more different aspects. Or we might resign ourselves to making claims only in probability – which will protect us from being disproved by any single event, but which may also increase immensely the overheads needed to justify precise quantitative assertions (of probabilities).

> *Tangles offer a structural, rather than probabilistic, way to afford the redundancy needed in such cases. They allow us to derive predictions from our data as we would expect them from identifying causes, while sidestepping the philosophical issue of what constitutes a cause.*

The idea, at a high level, is to replace the search for 'causes' with a search for something we can observe directly: structure in our data that occurs *as a result* of the presence of an underlying cause – a kind of structural footprint in observable data that we find whenever phenomena have some common cause, no matter what that cause may be. Different causes have different footprints, but all causes have the same *structural type* of footprint: tangles. The idea is that the structural footprint of each particular cause should carry enough information to replace any reference in the scientific process to that cause, e.g. in making predictions, with a reference to this structure, the tangle that reflects this cause in observable data. If desired, we may think of tangles as an extensional definition of 'cause' that achieves precision and observability at the expense of the intuitive appeal of our informal notion of 'cause'.

In our generic example, a tangle would be a set of *hypothetical* readings for the measurements we have performed on our phenomena, a set of one possible reading per measurement. It would not be just any such set, but one that is typical for the actual readings we got on our phenomena:

> *A tangle is a typical set of measurement readings, one for each measurement, such as a set of readings due to a particular cause.*

It may happen that one, or several, of our phenomena produced exactly this set of readings. But it can also happen that an 'abstract' set of readings is typical, and hence a tangle, for our collection of phenomena without occurring exactly in any one of them. This might be the case, for example, for a set τ of measurement readings produced by some given cause under laboratory conditions. This set of readings will be typical for our phenomena if they were indeed triggered by this cause, even if none of them produced exactly τ under our measurements.

What, then, does it mean that τ is typical for our phenomena? We shall address this question in detail below. Its most important aspect, however, is that our definition of 'typical' will *not* stipulate the existence of some common cause for our phenomena. It will be such that a single cause, or some fixed combination of causes, will produce a set of readings that satisfies our definition of 'typical'. But the definition itself will refer only to our data: the measurement readings we obtained on the phenomena we are seeking to explain. This will allow us to identify tangles directly in the data, without guessing at possible causes, and then *from* these tangles infer that, perhaps, some known cause is present.

Just as a set of similar phenomena can have several possible causes, our measurements might indicate the presence of a single tangle, or several, or none. Given any one of these tangles, we may try to find a common cause for this typical set of readings, or choose not to try. If there *is* a common cause for sufficiently many of the phenomena investigated, it will show up as a tangle and can thus be identified.

But there can also be tangles that cannot, or not yet, be 'explained' by a common cause. Such tangles are just as substantial, and potentially useful, as those that can be labelled by a known common cause; indeed perhaps more so, since the absence of an obvious common cause may have left them unidentified in the past. In this sense, identifying tangles in large sets of similar phenomena can lead to the discovery of new meta-phenomena that had previously gone unnoticed and might, henceforth, be interpreted as a 'cause' for the group of phenomena that gave rise to this tangle.

So when is a set τ of hypothetical measurement readings deemed 'typical' for the actual measurements taken on our phenomena, and is therefore a tangle? There are two notions of 'typical' that are important in tangle theory: a strong one that is satisfied by many tangles but not required in their definition, and a weaker one that is required in their definition, and which suffices to establish the main theorems about tangles.

The strong notion, which we might call *popularity-based*, is that our set of phenomena has a subset X – not too small – such that, for every measurement made, some healthy majority – more than two thirds – of the phenomena in X yielded the reading specified by τ for that measurement.[2] Note that these may be different two thirds of X for different measurements: every phenomenon, even one in X, may for *some* measurements produce readings different from the readings that τ specifies for those measurements. But every entry in τ reflects some aspect of what we measured that many of the phenomena in X have in common, and is in this sense typical for all the phenomena in X. Clearly, there can be several such tangles τ, witnessed by different sets X of phenomena.

The weaker notion of when our set τ of hypothetical measurement readings is 'typical' for the actual readings obtained from our measurements, and hence constitutes a tangle, might be called *consistency-based*. It requires of τ that, for every set of up to three of our measurements, at least one of the phenomena we measured gave the readings specified by τ for these three phenomena.[3] Note that if τ is typical in the popularity-based sense it will also be typical in this consistency-based sense,[4] but not conversely.

We often strengthen these consistency requirements of a tangle by asking a little more: that, for some 'agreement parameter' n we chose for the given context, every three measurement readings specified by τ must be shared not only by one of the phenomena we measured but by at least n of them. Such sets of hypothetical measurement readings, one for every measurement, will thus be even more typical for our phenomena, the more so the larger n is.

All three of these notions of 'typical' are robust against small changes in our data. This makes tangles well suited to 'fuzzy' data with the kind of inherent variation indicated earlier. But the definition of tangles will be completely precise: a formal description of the structure of our data *including* any aspects of its fuzziness. We shall therefore be able to use tangles in rigorous mathematical analysis of our data as it comes.

1.2 Tangles in the social sciences

Suppose we run a survey S of political questions on a population P of a thousand people. If there exists a group of, say, a hundred like-minded people among these, there should be a way of answering all the questions in S that is typical for those one hundred people: that, after all, is what 'like-minded' means. Let us write X for this set of a hundred people, and τ for the way of answering S that is typical for them. Thus, τ is one typical set of answers to all the questions in S.

Let us try to quantify this last assertion, that our answer set τ is typical for the people in X. One way to do this is to require that, for every question s in S, some healthy majority – more than two thirds, say – of the people in X agree with the answer to s that τ specifies. Note that *which* two thirds of X these are may differ for different questions s. Every answer specified by τ reflects the views of a large majority of the people in X, and is typical for all of X in this way. But we may not be able to pin down anyone in P, let alone two thirds of the people in X, that answered all of S as specified by τ.

We shall call such a complete collection τ of views that is typical for some of the people in P a *mindset tangle*. There may be more than one mindset tangle for S, or none, just as there may be several groups of like-minded people in P, or none, each with their own typical way of answering S.

Traditionally, mindsets are found intuitively: they are first guessed, and only then established by quantitative evidence, perhaps even from a survey designed specifically to test them. Mindset tangles can do that, too. For example, we might feel that there is a 'socialist' way σ of answering our survey S. We could write these answers down without looking at the actual returns for S, just appealing to our intuitive notion of what a 'socialist' way to answer S would be. To test our intuition that this is indeed a typical mindset for the people in P we might then check whether, in the actual returns for S we received from the people in P, we can find a sizeable subset X of P as earlier for this particular $\tau = \sigma$.

But tangles can also do the converse: we can identify mindsets, as tangles, in the returns for S without having to guess their answers first:

> *Tangles offer a precise and quantitative way to test for suspected mindsets in a population and to discover unknown ones.*

For example, tangle analysis of political polls in the UK in the years before the Brexit referendum might have detected the existence of an unknown mindset of voters across the familiar political spectrum that helped establish the surprise majority for Brexit in 2016. And similarly in the US with the MAGA[5] mindset before 2016, or that of conservative Greens in Europe in the early 1970s. Tangles can identify previously unknown patterns of coherent views or behaviour.

1.3 Tangles in data science

One of the most basic, and at the same time most elusive, tasks in the analysis of big datasets is *clustering*: given a large set of points in some space, one seeks to identify within this set a small number of subsets, called 'clusters', of points that are in some sense similar. If we visualize similarity as distance, clusters will be sets of points that are, somehow, close to each other.

Figure 1.1 shows a simple example of points in the plane. In the picture on the left we can clearly see four clusters. Or can we? If a cluster is a set of points that are pairwise close, and the two points shown in green in the right half of the picture lie in the same cluster, should not the two red points – which are much closer – lie in a common cluster too?

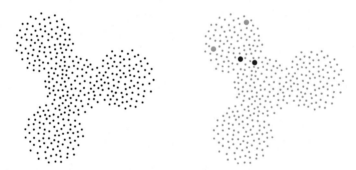

FIGURE 1.1.   Four clusters?

For reasons such as this, and other more subtle ones, there is no universal notion of cluster in data science. In our example there are 'clearly' four clusters – but it is hard to come up with an abstract definition of 'cluster' that is satisfied by exactly four sets of points in Figure 1.1, let alone four sets resembling those that we intuitively see as clusters.

Tangles seek to describe clusters in an entirely different manner. Not by dividing the dataset into subsets in some clever new way, but without dividing it up at all: although there will be four tangles in our picture, these will not be defined as sets of points. In particular, questions such as whether the green points should end up in the same cluster but the red points, perhaps, should not, do not even arise.

By avoiding the issue of assigning points to clusters altogether, tangles can be precise without making arbitrary and unwarranted choices:

Tangles offer a precise, if indirect, way to identify fuzzy clusters.

Rather than looking for dense clouds of data points, tangles look for the converse: for obvious 'bottlenecks' at which the dataset naturally splits in two. We call ways of splitting our dataset into two disjoint subsets *partitions* of the set, and the two subsets the *sides* of the partition.

Figure 1.2 shows three partitions of our point set at bottlenecks.[6] Now, whatever formal definition of 'cluster' one might choose to work with, one thing will be clear: no bottleneck partition will divide any cluster roughly in half, since that should violate either the definition of a cluster or that of a bottleneck. For example, given one of the three bottlenecks in our picture, and one of the four obvious clusters, we might argue over a few points about whether they should count as belonging to that cluster or not, or on which side of the bottleneck they lie. But for almost all the points in our picture these questions will have a clear answer once we consider a fixed cluster and a fixed bottleneck, no matter how exactly these may be defined.

Put another way, whichever precise definition of cluster (and of bottleneck) someone chose to work with, each of our four intuitive clusters would lie *mostly* on the same side of any partition at a bottleneck. Let us then say that the cluster *orients* this partition towards the side on which most of it lies. Figure 1.2 shows how the central cluster, no matter how it was defined precisely, orients the partitions at the three bottlenecks in

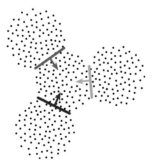

FIGURE 1.2.   Orienting the bottlenecks consistently towards the central cluster.

this way. Each of the four clusters assigns its own set of arrows to these same three partitions, and the central cluster orients them all inwards.

Note that assignments of arrows to bottlenecks that come from one of the four clusters in this way are not arbitrary: the arrows are consistent in that they all point roughly the same way, towards that cluster.

The key idea behind tangles, now, is to keep for each cluster exactly this information – how it orients all the bottleneck partitions – and to forget everything else (such as which points might belong to it). More precisely, tangles will be *defined* as such abstract objects: as *consistent orientations of all the bottleneck partitions* in a data set.

In this way, tangles will extract from the various explicit ways of defining clusters as point sets something like their common essence. Tangles will be robust against small changes in the data, just as they are robust against small changes in any explicit definition of point clusters that we might use to specify them. But their definition as such will be perfectly precise, and involve no arbitrary choices of the kind one invariably has to make when one tries to define clusters as sets of points.

To make this approach work, of course, one has to define formally what the 'bottleneck partitions' of a given dataset are, and when an orientation of all these bottleneck partitions is deemed to be 'consistent'. In our example of Figure 1.2 we defined both these with reference to those four intuitive clusters. Indeed, as 'bottleneck partitions' we took those that split the set of points where this appeared narrow in the picture, which is just another way of saying that we took precisely those partitions that did not cut right through any of the four intuitive clusters; and we called a way of orienting these three partitions 'consistent' if the arrows indicating this pointed towards one of those intuitive clusters.

If we are serious about defining tangles as abstract objects, however, in a bid to bypass the difficulties inherent in trying to define clusters as point sets, then this would beg the question. *The challenge is to define both bottleneck partitions and consistency of their orientations without reference to any perceived cluster*, however vaguely defined. Only once we have achieved this can we 'define' clusters not explicitly as point sets but indirectly as tangles, as is our aim.[7]

To make this challenge a little clearer, let us look at a slightly modified example. Figure 1.3 again shows four clusters with three bottlenecks. This time, one of these has an elongated shape, like a handle. As before, there are clearly four clusters, yet there is no obvious way to define them directly as point sets.

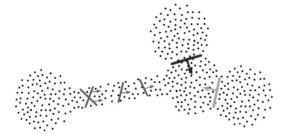

FIGURE 1.3. Three bottlenecks, but many bottleneck partitions,
all oriented consistently by the central cluster.

But now there appears to be a problem with our indirect approach
too. Intuitively, we would like to orient those bottlenecks; formally we
can orient partitions (at bottlenecks), by choosing one of their two sides;
but now there are many partitions 'at' the same bottleneck of the handle.
Which of these shall we choose to represent that bottleneck?

Since there is no canonical way to make this choice in the abstract,
we shall simply consider all those partitions at once: we shall work with
'bottleneck partitions' directly, and find ways to define these formally,
but we shall abandon our initial aim to formalize the more intuitive idea
of bottlenecks themselves. Our notion of consistency, then, will have to
apply also to how we orient partitions at the same bottleneck.

Let us see how this can work in Figure 1.3. We would like there to
be four clusters, no more, so only four of the many ways of orienting all
its bottleneck partitions should count as 'consistent'. In the picture this
can be achieved if, and only if, we can ensure that consistent orientations
of partitions at the same bottleneck always point the same way.[8]

In our example, the orientations of bottleneck partitions induced by
one of the four obvious clusters satisfy this nicely: any given cluster will
either lie mostly on the left of *every* partition at the handle, or mostly
on the right of every partition at the handle. Hence, the arrows defined
at these partitions by any of our four clusters will either all point to the
left, or they will all point to the right, and thus be intuitively consistent.

The challenge remains to come up with a formal definition of consistency as the basis for our notion of tangle that bears this out: one
that does not refer to any perceived clusters, but which in the above example will orient all the partitions at the handle in the same direction.
Chapter 2 shows how this can be done.

The task of identifying bottleneck partitions in a given dataset without reference to intuitively perceived clusters will be our topic later, in Chapter 9. Until then we shall usually assume a set of bottleneck partitions as given.

Once these two tasks are achieved, however, we shall have a definition of tangle which, while being entirely formal and precise, will be able to capture 'fuzzy' clusters in a robust way that does not require us to allocate points to clusters.

1.4 On the use of mathematical language in this book

In order to make tangles applicable in such diverse contexts as discussed in this chapter, the language in which we describe them can be based only on what all these contexts have in common. The basic language of modern mathematics is ideally suited to this. It uses only a few fundamental terms: sets and their elements, intersections and unions of sets, perhaps functions that assign to each element of one set some element of another set. And that's all we shall need for the start.

Since this basic mathematical language is so simple, it will be easy to ensure that all definitions and statements made in it will be reliable: they can be taken out of context, referred back to from any point later in the book, and should still stand up to scrutiny. Readers from outside mathematics may find this unusual, and are especially encouraged to rely on it: should they ever find themselves unsure of what some passage later in the book is intended to mean, the solution is likely not just to look out for redundancy near that passage, hoping it is repeated there in slightly different terms, but to look up the definitions of the terms involved.

Another important aspect of mathematical use of language is that, while all notions are precisely defined, the words we use to denote them are usually borrowed from everyday language – like 'tangle'. The reason is that inventing a new word for every new notion would soon make any text unreadable: there are just too many new notions invented all the time. To help with readability, moreover, we try to choose not just any ordinary word for some given new notion, but one that our memory can easily associate with it: a word whose everyday meaning resembles, a little at least, that new mathematical notion. When such a word is then encountered later in the text, it should already be flagged in the reader's mind as having a technical meaning, and it will usually be that meaning, not its everyday meaning, that is meant.

Still, there will remain cases where such ordinary words are still used with their ordinary meaning, and the reader will have to guess from the context when this is the case. A good example is the word 'typical', to which we assigned a technical meaning in Section 1.1 but which we used naively again at the start of Section 1.2. This was necessary, because the later section was meant to be readable as an entry point to this book. But any reader that happened to have read the earlier section first had to guess from the context at the start of Section 1.2 that 'typical' is used there with its everyday meaning.

While we usually hijack everyday words for mathematical notions, individual mathematical objects are often denoted by strange-looking symbols. These are chosen to help the reader remember how these objects relate to others, or how they might be interpreted when tangles are applied. For example, if we chose to denote a bottleneck partition as discussed in Section 1.3 by the letter s, it will be natural later to denote its two orientations as $\vec{s}$ and $\overleftarrow{s}$. This is innocuous enough. But remember that s, a partition of some dataset V, is a pair of subsets of V, its two *sides*. Now it may happen in our narrative that one of these sides occurs first, and that we wish to consider s only *because* it has this subset of V as a side. How, then, shall we refer to this side before s is mentioned?

One way to do this would be to introduce that side as a set denoted by some generic symbol for sets, such as A, then define B as its complement $V \setminus A$, introduce s as the set $\{A, B\}$, note that this is a partition of V with sides A and B, remember that orienting s means picking one of its sides, and finally define $\vec{s}$ as the 'orientation' A of s. Apart from the fact that this last step looks notationally odd, introducing the letters A and B is also a lot of notational clutter if all we ever need is s and its two orientations. In such cases we may therefore start our story by giving some interesting subset of V the name $\vec{s}$: just like that, out of the blue.

To non-mathematicians this may look strange, perhaps a little terrifying. But probably only because they suspect some hidden information in this notation which they think they do not know. When this happens, there is a very simple way to keep ones head: to rely on what exactly the text says, and not to worry about the notation. If $\vec{S}$ is introduced as an arbitrary finite set, then that's what it is: a set, nothing special. The reason for calling it $\vec{S}$ rather than A, say, may become apparent later,[9] but it will help not to speculate about this too much too early.

The whole point of using mathematical language is that its notions are not overloaded with meaning developed through the centuries, and

perhaps subject to debate. Starting our formal description in Part III
of what tangles are with just arbitrary sets, their elements and so on,
we shall have a clean slate that will be filled only carefully and slowly.
Every statement in the text should be understandable, if necessary, on
the basis of just the few things said formally before in Part III, so these
can be checked if in doubt.

Yet human readers are not machines, and sometimes it helps to have
an intuition, however vague, of what some formal notions are intended
to describe. Our formal development of tangle terminology will therefore
be accompanied by plenty of informal examples.

At the beginning, in Chapter 2, the narrative will in fact be almost
entirely in terms of examples. However, in order to become acquainted
early with the symbols to be used later in our more formal discourse,
we shall introduce these symbols here already: not in a mathematically
rigorous way as abstract sets, but naively as sets of 'things' that are
not mathematical objects. As the book progresses, I shall make more of
an effort to keep the levels of formal definitions and of examples visibly
separate: so that examples are not formally relied on later, while formal
definitions that are meant to be relied on can be easily be identified.

2

The notion of a tangle

The three introductions to tangles given in Chapter 1 focussed not so much on what tangles are but on what they might achieve: to offer a structural and evidence-based alternative to the vague notion of a 'cause' in the natural sciences and to identify these in any given data; a way of recognizing and discovering mindsets in the social sciences; and a new method of clustering in data science.

In this chapter we shift our focus to describing what tangles are. Although we shall frequently refer to the three introductions as background, we shall develop the notion of a tangle again from scratch. In particular, we shall not build on the various notions of 'typical' that we discussed in Section 1.1. Our exposition will still be informal; our mathematically rigorous treatment of tangles will begin in Part III.

Consider a collection V of objects and a set $\vec{S}$ of features[1] that each of the objects in V may have or fail to have. Given such a (possible) feature $\vec{s} \in \vec{S}$, we denote its negation by $\overleftarrow{s}$. The unordered pair $\{\vec{s}, \overleftarrow{s}\}$ of the feature together with its negation is then denoted by s, and the set of all these s is denoted by S. We may think of these $s = \{\vec{s}, \overleftarrow{s}\}$ as 'potential features': a feature bundled up with its negation into one entity.

For example, if V is a set of pieces of furniture, then $\vec{s}$ might be the feature of being made entirely of wood. Then $\overleftarrow{s}$ would be the feature of being made of any other material, or a combination of materials, and s could be thought of as the question of whether or not a given element of V is made entirely of wood.

In the language of Section 1.1 the elements of V would be the phenomena investigated. The $s \in S$ would be the measurements performed on these phenomena, with two possible outcomes $\vec{s}$ and $\overleftarrow{s}$ (called 'readings' in Section 1.1).

In the example of Section 1.2, the set V would be the population P of people polled by our survey S, which for simplicity we assume to consist of yes/no questions. If $\vec{s}$ denotes the 'yes' answer to a question $s \in S$, then $\overleftarrow{s}$ will denote the 'no' answer, and vice versa.

In the clustering scenario of Section 1.3, the set V would be the set of points in which we look for clusters. If we equate a feature $\vec{s}$ with the set of objects in V that have it, then $\vec{s}$ and $\overleftarrow{s}$ form a partition of V, the partition $s = \{\vec{s}, \overleftarrow{s}\}$. We may think of S as the set of those partitions of V that are particularly natural, its 'bottleneck' partitions.

2.1 Features that often occur together

Tangles are a way to formalize the notion that some features typically occur together. They offer a formal way of identifying such groups of features as 'typical for V', each 'type' giving rise to a separate tangle.

Before we make this more precise, let us point out right away that the term 'typical for V' will not normally be applied to single features, only to groups of features, which we then call *types*. We think of a group of features as 'typical for V' if these features often occur together: if the fact that a given $v \in V$ has one of them makes it more likely that it has the others too – but in a structural, not merely probabilistic, sense.

In order to identify a collection of features as 'typical', however, it will not be necessary to precisely delineate a corresponding set of *objects* (elements of V) that have exactly, or even mostly, these features. This reflects most real-world examples, where these sets are at best 'fuzzy'. By working directly on the level of features rather than the level of objects, tangles can be completely precise even when the objects whose features they capture cannot be clearly delineated from each other. This is a particular strength of tangles when they are used for clustering, as indicated in Section 1.3. However, tangles are qualitatively different; they are not just a new clustering method.

Let us return to the example where V is a set of pieces of furniture. Our list $\vec{S}$ of possible features (including their negations) consists of qualities such as colour, material, the number of legs, intended function, and so on – perhaps a hundred or so of possible features. The idea of tangles is that, even though $\vec{S}$ may be quite large, some of its elements may combine into groups[2] that correspond to just a few types of furniture as we know them: chairs, tables, beds and so on.

A key aspect of tangles is that they can identify such types without any prior intuition: if we are told that a container V full of furniture is waiting for us at customs in the harbour, and all we have is a list of items v identified only by an item number together with, for each number, a list of which of our 100 features this item has, our computer – if it knows tangles – may be able to tell us that our delivery contains furniture of just a few types: types that we (but not our computer) might identify as chairs, tables and beds, perhaps with the tables splitting into dining tables and desks.

In the language of Section 1.1 these types would correlate with the different possible 'causes' for objects to be furniture: our need to sit, sleep, eat and so on. In the example of Section 1.2 they would be mindsets. In the setting of Section 1.3, the sets of chairs, tables and beds would form clusters in V. These clusters might not be clearly delineated – for example, if our delivery contains a deckchair – but the types, groups of features that often occur together, would be precisely defined.

In the remainder of this chapter we shall not always make explicit reference to the three example scenarios from Chapter 1. But readers are encouraged to check for themselves what the various new terms mean in each of those contexts, to keep all three aspects alive as they build their intuition for tangles.

2.2 Consistency of features

To illustrate how our computer may be able to identify types of furniture from those feature lists without understanding them, let us briefly consider the inverse question: starting from a known type of furniture, such as chairs, how might this type be identifiable from the data if it was not known to be a type?

A possible answer, which will lead straight to the concept of tangles, is as follows. Each individual piece of furniture in our unknown delivery, $v \in V$ say, has some of the features from our list $\vec{S}$ but not others. It thereby *specifies* the elements s of S: as $\vec{s}$ if it has the feature $\vec{s}$, and as $\overleftarrow{s}$ otherwise. We say that every $v \in V$ defines a *specification* of S, a *choice for each $s \in S$ of either $\vec{s}$ or $\overleftarrow{s}$ but not both.*[3] We shall denote the particular specification of S defined by v as

$$v(S) := \{\, v(s) \mid s \in S \,\},$$

where $v(s) := \vec{s}$ if v specifies s as $\vec{s}$ and $v(s) := \overleftarrow{s}$ if v specifies s as $\overleftarrow{s}$.

Does every specification of S come from some $v \in V$ in this way? Certainly not. There will be no object in our delivery that is both made entirely of wood and also made entirely of steel. Therefore no $v \in V$ will specify both r as $\vec{r}$ rather than $\overleftarrow{r}$, and s as $\vec{s}$ rather than $\overleftarrow{s}$, when $\vec{r}$ and $\vec{s}$ stand for being made of wood or steel, respectively. Now, S has many (abstract) specifications that contain both $\vec{r}$ and $\vec{s}$. But none of these is defined, as $v(S)$, by any real piece v of furniture, because the features $\vec{r}$ and $\vec{s}$ are inconsistent.

Let us turn this manifestation in V of logical inconsistencies within $\vec{S}$ into an extensional definition of 'factual' inconsistency for specifications of S in terms of V. Let us call a specification of S *consistent* if it contains no inconsistent triple, where an *inconsistent triple* is a set of up to three[4] features that are not found together in any $v \in V$. Specifications of S that come from some $v \in V$ are clearly consistent, because every three features in $v(S)$ are shared at least by v. But S can also have consistent specifications that are not, as a whole, defined by any $v \in V$ as $v(S)$.[5]

Tangles will be specifications of S with certain properties that make them 'typical' for V. Consistency will be a minimum requirement for qualifying as typical.

2.3 From consistency to tangles

It is one of the fortes of tangles that they allow considerable freedom in the definition of what makes an entire specification of S 'typical' for V – freedom that can be used to tailor tangles precisely to the intended application. We shall discuss this in detail in Chapter 7. But we are already in a position to mention one of the most common ways of defining 'typical', which is just a strengthening of consistency.

To get a prior feel for our (forthcoming) formal definition of 'typical', consider the specification of S in our furniture example that is determined by an 'ideal chair' plucked straight from Plato's heaven: let us specify each $s \in S$ as $\vec{s}$ if this imagined ideal chair has the feature $\vec{s}$, and as $\overleftarrow{s}$ if not.[6] This can be done independently of our delivery V, just from our intuitive notion of what chairs are. But if our delivery has a sizeable portion of chairs in it, then this phantom specification of S that describes our ideal chair has something to do with V after all.

Indeed, for every triple $\vec{r}$, $\vec{s}$, $\vec{t}$ of features of our ideal chair there will be a few elements of V, at least n say, that share these three features. For example, if $\vec{r}$, $\vec{s}$, $\vec{t}$ stand for having four legs, a flat central surface, and a near-vertical surface, respectively, there will be – among the many chairs in V which we assume to exist – a few that have four legs and a flat seating surface and a nearly vertical back.

By contrast, if we pick twenty rather than three features of our ideal chair there may be no $v \in V$ that has all of those, even though there are plenty of chairs in V. But for every choice of three features there will be several – though which these are will depend on which three features of our ideal chair we have in mind.

Simple though it may seem, it turns out that for most sensibly imagined furniture deliveries and reasonable lists S of potential features this formal criterion for 'typical' distinguishes those specifications of S that describe genuine types of furniture from most of its other specifications.[7] But in identifying such specifications as 'types' we made no appeal to our intuition, or to the meaning of their features.[8]

So let us make this property of specifications of S that describe 'ideal' chairs, tables or beds into a more formal, if still ad hoc, definition of 'typical': let us call a specification τ of S *typical* for V if for every set R of up to three elements of S there are at least n elements v of V that specify R as τ does, i.e., which satisfy $v(s) = \tau(s)$ for every $s \in R$.[9] The integer n here is a fixed parameter on which our notion of 'typical' depends, and which we are free to choose. We allow $n = 1$, in which case 'typical' means no more than 'consistent'. But as we make n larger, the resulting notion of when a specification of S is 'typical for V' gets stronger and stronger.

Crucially, this definition of 'typical' is purely extensional: it makes no reference to what a typical specification of S is 'typical of'. Specifications of ideal chairs, tables or beds are all typical in this same sense: they all satisfy the same one definition.

Equally crucially, a specification of S can be typical for V even if V has no element that has all its features at once. Thus, we have a valid and meaningful formal definition of an 'ideal something' even when such a thing does not exist in the real world, let alone in V.

Relative to our notion of 'typical', which is subject to change as we increase n, we can now define tangles informally:

A tangle of S is any specification of S that is typical for V.

Since our ad hoc definition of 'typical' is phrased in terms of small subsets of $\vec{S}$, sets of size at most 3 (of which there are not so many), we can compute tangles without having to guess them first. In particular, we can compute tangles of S even when V is 'known' only in the mechanical sense of data being available (but not necessarily understood), and S is a set of potential features that are known, or assumed, to be relevant but whose relationship to each other is unknown.

Tangles therefore enable us to *find* even previously unknown 'types' of features in the data to be analysed: combinations of features that occur together significantly more often than others. This was important in all three of the scenarios from Chapter 1: tangles can identify previously unknown causes, mindsets, or clusters.

When we treat tangles mathematically in Part III, we define them more broadly as consistent specifications of S that have no subset in some collection $\mathcal{F}$ of subsets of $\vec{S}$, which we may specify as we wish. If we choose $\mathcal{F} = \emptyset$, then all consistent specifications of S will be tangles. Our informal definition of 'typical' then corresponds to choosing $\mathcal{F}$ as

$$\mathcal{F}_n := \big\{ \{\vec{r}, \vec{s}, \vec{t}\} \subseteq \vec{S} : |\vec{r} \cap \vec{s} \cap \vec{t}| < n \big\},$$

where $|\ |$ denotes the number of elements of a set and the features $\vec{r}, \vec{s}, \vec{t}$ are interpreted as the subsets of V that have them. So their intersection is precisely the set of all $v \in V$ that have all three features $\vec{r}, \vec{s}$ and $\vec{t}$.[10]

A tangle of S, then, with 'typical' defined as earlier, or more formally as having no subset in $\mathcal{F} = \mathcal{F}_n$, is any specification of S such that every three of its features are shared by at least n elements of V. We call this n the *agreement* value required of these tangles; the variable n in the definition of $\mathcal{F}_n$ is its *agreement parameter*. In general, a tangle of S will be any consistent specification of S no subset of which is an element of $\mathcal{F}$, for any collection $\mathcal{F}$ of subsets of $\vec{S}$ of our choice.

With this notion of a tangle, any reader who cannot wait to see some potential tangle applications is well equipped to skip straight ahead to Part II now, which requires no more knowledge about tangles than their definition. The remainder of Part I offers an introduction to what tangle theory has to offer, which is treated more formally in Part III and will form the basis of our more detailed look at tangle applications in Part IV.

2.4 Principal and black hole tangles: two simple examples

In this section we introduce two simple, if somewhat extreme, examples of tangles that will crop up, again and again, throughout the book – often as tangles we want to watch out for because they are not the kind of tangles we are really interested in.

Assume first that, as in Section 1.3, our set S of potential features consists of partitions of V into two non-empty subsets. Every feature $\vec{s}$, then, is one of the two sides of a partition $\{A, B\}$ of V: either A or B. If $\vec{s} = A$ then $\overleftarrow{s} = B$, and vice versa. In particular, every feature is formally a subset of V, and every specification of S, including any tangle, will be a set of subsets of V.

As our first example, consider for any given $v \in V$ the specification

$$\tau_v = \{\, \vec{s} \in \vec{S} \mid v \in \vec{s} \,\} = v(S)$$

of S. These are the *principal* specifications of S, one for every $v \in V$. Note that they are consistent, since every $\vec{s} \in \tau_v$ contains v.

If $\mathcal{F} = \emptyset$ in our formalization of tangles at the end of Section 2.3, these consistent specifications τ_v of S will be tangles; we call them the *principal tangles* of S. Let us look at a particularly common example.

We say that a specification σ of S is *focussed on* $v \in V$ if it contains the singleton subset $\{v\}$ of V as an element. If σ is consistent, then this implies that $\sigma = \tau_v$: given any $\vec{s} \in \vec{S}$ such that $v \in \vec{s}$ (as in the definition of τ_v), we cannot have $\overleftarrow{s} \in \sigma$, since $\overleftarrow{s}$ is inconsistent with $\{v\} \in \sigma$, so $\vec{s} \in \sigma$. Hence σ specifies every $s \in S$ as τ_v does, which means that $\sigma = \tau_v$.

Thus, if S has a tangle focussed on some $v \in V$, then this is its principal tangle τ_v. Conversely, τ_v need not be focussed on v, even if it is a tangle, because $\{v\}$ may fail to be an element of $\vec{S}$. But τ_v cannot be focussed on any $u \neq v$ either, since $\{u\} \notin \tau_v$ by definition of τ_v.

Let us show that these τ_v are the *only* consistent specifications of S, and hence its only possible tangles, when S is not a set of particularly natural 'bottleneck partitions' of V but consists of all the partitions of V. Since $\{\{v\}, V \smallsetminus \{v\}\}$ lies in this S, the principal specification τ_v of S will then contain $\{v\}$, and hence be focussed on v.

To prove that the τ_v are the only consistent specifications of the set S of *all* partitions of V, consider any such specification τ. Let $\vec{s} = A$ be a smallest element of τ in terms of $|A|$, the number of elements of V it contains. Note that $A \neq \emptyset$: being consistent, τ cannot contain the set

$\{A, A, A\} = \{A\}$ if the intersection of its elements is empty, which it is
if $A = \emptyset$. Let us show that $|A| = 1$.

Suppose A has more than one element. Then A has a partition into
two non-empty subsets, B and C say. By the minimal choice of A these
do not lie in τ, so their complements $\vec{r} := V \smallsetminus B$ and $\vec{t} := V \smallsetminus C$ do. But
now $\{\vec{s}, \vec{r}, \vec{t}\} \subseteq \tau$ while $\vec{s} \cap \vec{r} \cap \vec{t} = \emptyset$, contradicting the consistency
of τ. Hence our assumption that A has more than one element is false:
it consists of a single element, v say.

Let us prove that $\tau = \tau_v$. Given any partition of V, one of its two
sides must be in τ. In fact it must be the side containing v, since the
other side has empty intersection with $\{v\}$, so they cannot both lie in τ.
Hence τ specifies every partition in S as v does: $\tau(s) = v(s)$ for every
$s \in S$. This completes our proof that $\tau = \tau_v$.

Depending on our choice of $\mathcal{F}$, this has the following consequences
for tangles of this extreme choice of S. If $\mathcal{F} = \emptyset$, the tangles of S are
precisely its consistent orientations, those of the form τ_v. As soon as we
forbid singletons $\vec{s} = \{v\}$ as elements of tangles, however, e.g. by taking
$\mathcal{F} = \mathcal{F}_n$ with $n > 1$, or by explicitly putting all sets $\{\{v\}\}$ in $\mathcal{F}$, we have
no tangles of S at all: since tangles have to be consistent, they can only
be of the form τ_v, but we have just ruled those out.

For this reason, we shall not normally consider as S the set of all
partitions of V, but mostly sets of partitions that divide V in a particularly natural way. Those will be our 'bottleneck' partitions. Exactly
which these should be will be a matter for Chapter 9.

Incidentally, our earlier proof – not the fact – that the set of all
partitions of V has only the τ_v as consistent orientations can teach us
something we left unproved in Section 1.3: that any tangle τ of the
bottleneck partitions in Figure 1.3 orients all the partitions at the handle
in the same direction. The proof is very similar.[11]

Let us now look at the other extreme: that S contains too few partitions of V – e.g., because the 'bottlenecks' we are allowing are too narrow.
For example, let us assume that the only partitions in S are those of the
form $\{\{v\}, V \smallsetminus \{v\}\}$, and let τ be any consistent specification of S.

If τ contains one of the singleton sets in $\vec{S}$, say $\{v\} \in \tau$, then by
consistency it cannot contain another such set $\{u\}$ with $u \neq v$, since
$\{u\} \cap \{v\} = \emptyset$. It will therefore contain all the sets $V \smallsetminus \{u\}$ for such u.
All these contain v, so $\tau = \tau_v$. On the other hand, as long as $|V| \geqslant 4$,
also $\tau = \{V \smallsetminus \{v\} : v \in V\}$ is a consistent specification of S, a rather
'unfocussed' tangle of S.

If we expand the singleton sets $\{v\}$ to larger subsets of V which, intuitively, form 'clusters' of our data, we can generalize this last example to slightly more general examples of the same type, which we shall revisit several times later in this book: the *black hole tangles*.[12]

To define these more formally, let V have $n \geqslant 4$ disjoint clusters, well separated from each other. Let S consist of only the n partitions of V that each have one of the clusters on one side and the other $n-1$ clusters on the other side. Each of the clusters induces a tangle of S: the specification of S which, in the language of Section 1.3, orients every partition $s \in S$ towards that cluster. But there is one more tangle: that which orients each of the n partitions in S away from the cluster it separates from the others. This tangle does not correspond to any cluster in V: like a black hole at the centre of a galaxy it sits at its void centre, with all the other tangles arranged symmetrically around it (Figure 2.1).

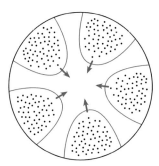

FIGURE 2.1.   The *black hole tangle* corresponds to no cluster.

2.5 Guiding sets and functions

When we defined a *tangle* of S, in Section 2.3, as any specification τ of S that is typical for V, we were assuming a notion of 'typical' that we called *consistency-based* in Section 1.1: for every set R of up to three elements of S there should be at least n elements of V that specify R as τ does, for some fixed $n \geqslant 1$ of our choice. In the more general definition of tangles at the end of Section 2.3, these tangles were obtained by setting $\mathcal{F} = \mathcal{F}_n$. These are the tangles we consider in this section, with $n = 1$ to allow the most general such tangles.

In Section 1.1 we also discussed another possible notion of 'typical', which we called *popularity-based*. This was that V has a subset X, not too small, in which τ is 'popular' in that for every $s \in S$ more than two thirds of the elements of X specify s as τ does. We saw that this implies that τ is typical also in the earlier sense, and hence is a tangle. We may think of X as 'witnessing' this, and as 'guiding' the choices made by τ.

Formally, let us say that a set $X \subseteq V$ *witnesses* that a specification τ of S is a tangle if every three features in τ are shared by some $x \in X$. If there are at least n such x for every three features in τ, we say that X *witnesses* that τ is a tangle *with agreement at least n.*

Let us say that X *guides* the specification τ of S if, for every $s \in S$, there are more x in X that specify s as τ does than there are $x \in X$ that specify s in the opposite way. If these majorities are greater than $2/3$, then X will also witness that τ is a tangle, no matter how large or small X is. The maximum $p > \frac{1}{2}$ such that, for every $s \in S$, at least $p|X|$ of the elements of X specify s as τ does, is the *reliability* of X as a guiding set, or *guide*, for τ.

In our furniture example, the tangle τ of being a chair is witnessed and guided by the set X of chairs in V if every feature of our 'ideal chair' τ is shared by more than two thirds of the set X of all the chairs in V. Such guiding sets X were also used in Section 1.2, where we defined a *mindset* as a collection of views established by a political survey S that were 'often held together', in exactly this sense.

More generally, let us say that a 'weight' function $w \colon V \to \mathbb{N}$ *guides* τ if, for every $s \in S$, the collective weight of all the $v \in V$ that specify s as τ does exceeds the collective weight of all the $v \in V$ that specify s in the opposite way.[13] If $X \subseteq V$ guides τ, then mapping the elements of X to 1 and all others to 0 is an example of a function that guides τ.

Much of the attraction and usefulness of tangles stems from the fact that, in practice, most of them have such guiding sets or functions [19]. But it is important to remember that the definition of a tangle does not require that such sets or functions exist. It relies only on the notion of consistency, which is defined with reference only to the values of $v(s)$ for the various $v \in V$ and $s \in S$.

In some contexts, however, tangles of S can be defined without any reference to V at all. In our furniture example we could have defined the consistency of a set of features, or predicates, about the elements of V in purely logical or linguistic terms that make no appeal to V. Indeed if $\vec{r}$ stands for 'made entirely of wood' and $\vec{s}$ stands for 'made entirely of steel', we said that the set $\{\vec{r}, \vec{s}\}$ is inconsistent. Our definition of this was extensional: that no object in V is made entirely of wood and also made entirely of steel. But we might have appealed instead to the fact that these two predicates are logically inconsistent – which implies that there is no such object in V, but which can be established without examining V.

The way consistency is defined in general [9], as part of the notion of abstract tangles, is something half-way between these two options: it makes no reference to V but refers only to some axiomatic properties of $\vec{S}$ which reflect our notion that $\vec{S}$ is a set of 'features'. In this way it also avoids any appeal to logic or meaning. Such abstract tangles, however, will not be treated in this book.

For our purposes the only important thing to note about guiding sets or functions is that while many tangles have them, tangles can be identified, distinguished, or ruled out without any reference to such sets or functions. The mindset of being socialist can be identified without having to find any actual socialists, let alone delineating these as a group of people against others. We shall return to guiding sets and functions in Sections 6.1, 14.1, and 14.2. The question of which tangles have guiding sets or functions is studied mathematically in [10] and [19].

3
The two main tangle theorems: an informal preview

Among the many mathematical theorems about tangles there are two fundamental ones that stand out. In terms of potential tangle applications, the first of these shows how tangles structure the dataset of which they are tangles. The second tells us how our data is structured if it has no tangle. Most tangle applications that go beyond finding tangles are based on one of these two results.

Continuing our informal approach from Chapter 2, let V again be a set of objects and $\vec{S}$ a set of possible features of its elements. Tangles are typical specifications of S, as defined in Section 2.3.

In order for the two tangle theorems to hold, the set $\vec{S}$ has to be rich enough. If it is not, we can make it so by adding to it some combinations of features already in $\vec{S}$. This will be explained in Chapter 7, Section 7.5. In this chapter we assume that $\vec{S}$ is rich enough in the sense required.

3.1 The first tangle theorem: how tangles structure our data

The first main tangle theorem finds a small subset T of S that suffices to distinguish all the tangles of S. Recall that every tangle of S is a (typical) *specification* of S, a choice of either $\vec{s}$ or $\overleftarrow{s}$ for every $s \in S$. Two different tangles of S, therefore, will make different such choices for at least one s – otherwise they would be the same tangle – and we say that every such s *distinguishes* these two tangles. The first of the two main tangle theorems says that we can always find such an s in some small set T which it extracts from S.

In the scenario of Section 1.1, this T will be a small set of particular measurements which, between them, suffice to distinguish all the tangles found in the data: for any two types of phenomena, or two causes of similar phenomena, there will be a measurement in T for which these specify different readings.

An expert system built to establish which illness is causing a patient to feel unwell[1] might start by asking for measurements from this set T. Since the measurements in T distinguish between all the illnesses, there will be a unique illness compatible with the readings obtained for the measurements in T. Moreover, T is, or can be chosen to be, minimal with this property. Once this unique illness has been identified as the likely cause of the patient's ailments, this diagnostic hypothesis can be tested by further measurements particularly relevant to this illness.

In the social sciences scenario of Section 1.2, where S was a survey[2] we have conducted, the set T would be a set of critical questions that suffice to distinguish the mindsets that exist in the population surveyed. For every two mindsets there will be a question in T on which they disagree, and which can thus be used to distinguish them. As an immediate application, T would make a good small questionnaire for a larger study if S was a pilot study designed to test a large number of questions on a smaller population.

In the clustering scenario of Section 1.3 the set T will be a set of partitions, at least one at each bottleneck. For every pair of tangles of the set S of all the bottleneck partitions there will be a partition in T which these two tangles orient differently.

3.2 The second tangle theorem: structure when there is no tangle

The second main tangle theorem tells us how our data is structured if it contains no tangle. The non-existence of a tangle in a dataset is a highly relevant and substantial piece of information, which tells us that the data is inconclusive in some objective and verifiable quantitative sense. In the mindsets scenario of Section 1.2, for example, the theorem furnishes inconclusive poll returns with a verifiable proof that mindsets not only were not found but do not in fact exist.

Given a set S that admits no tangle, the theorem returns a small subset T of S that already has no tangle: then S cannot have one either, since it would include one of T. Moreover, the theorem produces a small set of 'witnesses' to the non-existence of a tangle of T, and hence of S: a set of no more than $|T| + 1$ triples in $\vec{T}$ that are either inconsistent or in $\mathcal{F}$ (both of which a computer can easily check), and hence cannot be subsets of any tangle of T. These triples witness the non-existence of a tangle of T because they cover $\vec{T}$ in such a clever way – which a computer can check even without looking at V – that any tangle of T *would* have to contain one of these triples.

As with the first main tangle theorem, the set T and those triples cannot simply be found by trial and error. In both theorems, they are highly valuable in both senses of the word: knowing them gives us a lot of useful information that we would not otherwise have, and proving their existence requires some nontrivial mathematics. However, now that this has been done, they can often be found quickly by a computer, using the algorithms we shall meet in Chapter 11.

3.3 The predictive power of tangles

The general idea of trying to predict a person's likely behaviour in a future situation from observations of their actual behaviour in some past situations is as old as humanity: as we learn 'how someone ticks', we are better able to make such predictions.[3]

If we have the chance to choose, or even design, those earlier situations, e.g. by selecting them carefully from a collection of past situations in which our individual's behaviour was observed, or by devising a test consisting of hypothetical situations the response to which our individual is willing to share with us, we have a chance of improving our predictions by choosing such past or hypothetical scenarios that are particularly relevant. Tangles can help to identify these.

Let P be a set of people in whose likely actions we are interested, and $\vec{S}$ the set of possible such actions. To keep things simple, let us consider the example from the introduction in Section 1.2: there, $\vec{S}$ is a set of possible views of the people in P,[4] and we wish to predict how a person from P would answer the questions from S. Let us assume that, as the basis for our prediction, we are allowed to quiz that person on some small set T of questions, and base our prediction for their answers to S on their known answers to the questions in T. Our aim is to choose T so as to make these predictions as good as possible.

The first of our two main tangle theorems, discussed in Section 3.1, is designed to produce a particularly suitable set T of questions, one whose answers entail more predictive power for S than an arbitrary subset of S of that size would. This is because T consists of just enough questions to distinguish all the tangles of S: the typical ways to answer S, the mindsets (regarding S) that exist in P.

The answers of an individual to the questions in T thus determine exactly one such type, or mindset: there is only one typical way of answering all the questions in S that includes these particular answers to T. This typical way of answering S, a tangle of S, is a an especially good prediction for the answers of our individual to the other questions in S if we assume that he or she follows any mindset at all, which seems more likely than not.

On the face of it, there appears to be a problem with this approach in that we are trying to base our predictions for somebody's answers to the questions in S on being able to compute the tangles of S first, which in turn requires that we already know what we are trying to predict: the answers of the people in P to the questions in S. However, this is not in fact a problem. In any real-world application we would compute these tangles based on how the people in some representative subset V of P answered S, and then test the individual we are interested in, most likely someone in $P \smallsetminus V$, on the set T computed for these tangles.

In a typical application in the social sciences, S might be a pilot study run on a small subset V of a larger population P. Then $T \subseteq S$, once computed from the return of the pilot study, might be used as the main study run on P, with fewer but particularly relevant questions selected from S. The answers to T of an individual in P can justify predictions on how this individual would answer the rest of S, and T is a particularly well chosen subset of S for this purpose.

Part II

Tangles in Different Contexts

A collection of informal examples

The aim of the three chapters in this Part II of the book is to indicate the range of potential uses of tangles by a diverse collection of simple synthetic examples from different contexts. Our aim will be to flesh out the notion of a tangle, as developed in Part I, by viewing it in these contexts, not (yet) to discuss what applying tangles can achieve there.

After a more rigorous introduction to the mathematics of tangles in Part III, we shall return to these examples in Part IV. We will then be able to say a little more than we can now: what the tangle theorems from Part III mean in the contexts of these examples, and how they might be applied there.

Our presentation here will be at the level of informality adopted in Part I, and in the language introduced there. The sections can be read independently of each other. Each of them is prerequisite reading for the corresponding section in Part IV, but not for any other material in this book.

Since I am not an expert in any of the disciplines from which the examples described in the next three chapters are taken, I shall not attempt more than to indicate the kind of use that tangles might find there: any actual use will require genuine expertise in the respective field. However, the examples discussed here may serve as templates for such an endeavour: they should be diverse enough to suit most contexts, their description simple enough to be inspiring.

The default format for describing those examples will be to

- name the set V of objects studied;
- describe the set S of their potential features;
- see what tangles of S mean in this specific context.

In Section 2.3 we defined a *tangle* of S as a specification of S that is 'typical for V'. The minimum requirement for this was consistency: that any three features in the tangle be shared by some element of V. In the examples discussed in this chapter we shall sometimes make additional requirements on the tangles we consider. These will always be made in the same format: we may require that our tangles shall not contain certain subsets of $\vec{S}$. We then write $\mathcal{F}$ for the set of those 'forbidden subsets', and consider as tangles only those specifications of S that are consistent and, in addition, have no subset that is an element of $\mathcal{F}$.

A frequent choice for $\mathcal{F}$ is the set $\mathcal{F}_n$ of triples of features such that fewer than n elements of V have these particular (up to three) features, where n is an 'agreement parameter' we are free to set, perhaps relative to $|V|$. Tangles for this $\mathcal{F} = \mathcal{F}_n$ are specifications of S that are typical for V in the stronger sense (than mere consistency) that every three features in the tangle must be common to at least n elements of V. We may also consider collections $\mathcal{F}$ of forbidden subsets that include $\mathcal{F}_n$ for some n and contain some other subsets of $\vec{S}$ as well.

4

Examples from the natural sciences

This chapter is written in the informal terminology of Chapter 2. We have a set V of objects (called 'phenomena' in Section 1.1), and a set $\vec{S}$ of possible features of these objects. These features come in *inverse* pairs $\{\vec{s}, \overleftarrow{s}\}$, which we call 'potential features' and denote by dropping those opposite arrows, in our case by s. The inverse features $\vec{s}$ and $\overleftarrow{s}$ are the two *specifications* of $s = \{\vec{s}, \overleftarrow{s}\}$. Every object $v \in V$ has exactly one of the features $\vec{s}, \overleftarrow{s}$ for every s; we denote the one it has by $v(s)$, and say that v *specifies* s as $v(s)$. We write S for the set of all s with $\vec{s} \in \vec{S}$.

In our natural sciences context, a potential feature s typically corresponds to a type of measurement that can be applied to the objects in V, and which has two possible outcomes: $\vec{s}$ and $\overleftarrow{s}$.[1]

A *specification* of S is a subset of $\vec{S}$ that contains either $\vec{s}$ or $\overleftarrow{s}$ for each $s \in S$, but never both. It is *consistent* if for every three of its choices, $\vec{r}, \vec{s}, \vec{t}$ say, there exists an object $v \in V$ that specifies r, s and t in exactly this way, as $v(r) = \vec{r}$ and $v(s) = \vec{s}$ and $v(t) = \vec{t}$. A consistent specification of S is a *tangle* of S if it has no subset that is an element of some set $\mathcal{F}$ of subsets of $\vec{S}$ that we may specify. If no such $\mathcal{F}$ is specified it will be $\mathcal{F} = \mathcal{F}_n$ for some n, with $n = 1$ as the default. When $\mathcal{F} = \mathcal{F}_1$, all consistent orientations of S are tangles.

Of the three examples of potential tangle applications that we discuss in this chapter, the first is the most generic: it illustrates how tangles can be used to build an *expert system*: a computer program that has access to a vast amount of expert information in some field, and aims to assist a professional in that field in a process of making decisions. The idea is that these decisions should be backed up by the expert information available, even when this body of information is too large to be known explicitly to any one person. The expert system will thus listen to the expert, but it will also actively make suggestions about which decisions to take next. We shall describe the possible use of tangles for building such a system by focussing on the process of medical diagnosis. This is not only particularly important in that it affects us all; it is also particularly suitable for tangle applications, because tangles are designed to see a clear picture when the quantitative information available is prone to a degree of imprecision or error.

The next example we will discuss is from molecular biology: how tangles might be used in DNA or protein sequencing. The elements in a sequence of nucleotides or amino acids are viewed as its features, and features that often occur together will form a tangle.

Our final example asks how tangles might be used to identify groups of features of pathogens that often occur together, and which thus suggest the development of a drug that can target all pathogens with these features at once.

All three examples will be revisited in Chapter 12, when we discuss them more formally on the basis of the tangle theory we will have met by then.

4.1 Expert systems: what is an illness?

What exactly is an illness? Since the advent of modern medicine, we have come to think of illnesses in terms of causes: a common cold, for example, is an infection by a certain type of virus.

However, notions of illnesses usually predate the discovery of their cause. In those cases, illnesses are defined in terms of what we see, no matter whether we understand it: as collections of symptoms. Such definitions, by nature, are much vaguer. As medical research advances, shall we seek to abandon these symptomatic definitions of illnesses in favour of naming one, or perhaps a definitive combination of a few, measurable conditions whose presence defines an illness as what they cause? Or shall we stick with the symptomatic definitions – and if so, is there anything we can do to handle their intrinsic vagueness in precise, quantifiable, and intersubjective ways?

Even from a modern perspective, illnesses do not normally have a single cause. This might still be argued in rare cases, such as a missing gene. But even an infection by a virus that is often behind the common cold turns into an illness only if other 'causes' are present too, such as a weakened immune system. And besides, even if a single cause, or definitive set of causes, exists, this will rarely be what we observe: what we see is a patient who feels unwell. Such a patient will come with some obvious symptoms, and can be tested for some less obvious ones. Doing this before making a wild guess at a single cause is a time-tested, and still entirely sensible, process of diagnosis.

Doctors like to speak of individual tests in this process as 'excluding' some potential cause. The idea is that any given hypothetical cause of the illness would also cause certain measurable others conditions, such as body temperature in a certain range, so that if the measured temperature lies outside this range then this hypothetical cause cannot be the cause of the illness. As we all know, while this makes sense in principle, it does not work in every case: humans are just too complex, and the condition whose measurement we hope might exclude our potential cause is likely also influenced by a host of other parameters beyond our control.[2]

What doctors will do in such cases is work with intuitive probabilities: while each symptom tested may sometimes fail to be present as expected, the chance that this fails for many independent symptoms is much smaller, and so the diagnostic conclusion drawn (such as the exclusion of a certain cause) becomes more certain. An expert system should then advise them on just what the probabilities are – a pretty daunting task considering the number of experiments it takes to make probabilistic statements with any reasonable degree of confidence.

Tangles cope with the intrinsic fuzziness in how illnesses are reflected by symptoms in a structural, rather than probabilistic, way. They can serve to identify illnesses from sets of diagnostic measurements, and they can help identify combinations of diagnostic results as 'typical' – in which case these could, and maybe should, be thought of as a previously unknown illness that doctors may wish to be aware of in the future.

Let us now have a look at how this might work in the setting of Chapter 2: how illnesses correspond to tangles. The first of the two main tangle theorems, applied to these tangles, will find what in our context amounts to an expert system for the diagnostic process of identifying an illness: a small set T of measurements which, between them, suffice to distinguish different illnesses encoded as tangles, with minimum redundancy (Chapter 12, Section 12.2). Once a particular tangle has been identified by the system, it can then be verified further by additional measurements of the symptoms that define it.

So how can an illness be described as a tangle? In our model, V might be the set of all patients on record. The set S might consist of measurements, ideally (for simplification) with just two potentials results: $\vec{s}$ for a 'positive' reading indicating an abnormality (such as 'increased pulse'), and $\overleftarrow{s}$ for the corresponding 'negative' reading ('pulse slow or normal').[3]

Any illness then defines a specification of S, one that specifies those measurements $s \in S$ as $\vec{s}$ that usually, though not necessarily always, have a positive reading when the illness is present, and all other s as $\overleftarrow{s}$. Such a specification τ of S will be 'typical' in both senses discussed in Section 1.1, and will hence be a tangle.[4]

Conversely, every tangle of S is a collection of symptoms that often occur together. This alone justifies giving this collection of symptoms some attention, even if it is not normally associated with any known common illness. And it might be worth thinking of it as an illness in the future: it is not uncommon in medicine that medical conditions that previously escaped attention become recognized as illnesses. Tangles are ideal to help with this process, because they identify just this: collections of symptoms that often occur together.

More broadly, tangles might be useful for expert systems in other diagnostics contexts too, not only of machine failures or computer crashes. As artificial intelligence advances and can be applied to extract answers to factual questions from texts even if these did not address those questions explicitly, or answers to content-related questions about scenes shown in a picture, we can also treat such questions as 'measurements', or indeed as questionnaires, on texts and images. We might then compute their tangles to build expert systems, say, for the classification of paintings in art history, or even to identify types of similar historical developments that might teach us lessons for the present.

4.2 DNA tangles: identifying organisms from imperfect data

In this example, V is a set of DNA molecules. Our aim is to use tangles to determine which species, subspecies, or individual organisms they represent. In Chapter 12, Section 12.3, we shall complement the method discussed here by a clustering-based method also using tangles. Either method will allow us to determine the phylogenetic tree of the organisms in our sample in a new way, based on the tangle theorems from Chapter 8.

In the same way one can classify proteins in terms of their sequences of amino acids. We consider DNA sequences here just to illustrate the ideas in how to use tangles for such analysis, at a level that assumes no expert knowledge. To simplify matters, we assume that the genome of the organisms we consider is encoded in a single DNA molecule.

In each of our molecules we consider both of its two strands of nucleotides with a default orientation.[5] The nucleotides in each strand then have well-defined *positions*, indicated by integers, and at every position we have one nucleotide. Each of these has one of four possible bases in it, adenine, thymine, guanine or cytosine, which we denote by their first letters A, T, G and C. Every strand thus gives rise to a sequence of these bases, one at every position. The two sequences coming from the opposite strands of the same molecule can be obtained from each other by replacing each base with its partner base and reversing the direction of the sequence; we call such a pair of sequences *inverse* to each other.

Very crudely, when we have two identical base sequences obtained from our sample, they will most likely come from the same organism. If we have two similar sequences, they may come from different organisms of the same species. Generally, the more similar two sequences are, the more closely related are the organisms from which they were taken.

Since for every base sequence we also have its inverse sequence in our sample – remember that we took both strands from each molecule – the converse implications to those above will hold only up to inverting sequences: two sequences from the same organism should be identical up to inversion,[6] two sequences from organisms of the same species should be similar either to each other or to each other's inverse, and so on.

Not all the positions will be relevant for distinguishing the organisms or species which our molecules represent, and we consider only those positions that are. Let us think of V as the multiset[7] of base sequences we obtained from our DNA molecules trimmed to these relevant positions.

The characteristic features of the elements v of V are which base is present at each position: we can reconstruct the DNA molecule that gave rise to v from this information. These features thus have the form $\vec{s} = \vec{s}(p, B)$ to tell us that in position p we find the base $B \in \{A, T, G, C\}$. By our convention, the feature $\overleftarrow{s} = \overleftarrow{s}(p, B)$ then encodes the information that the base at position p is not B but one of the other three bases.

What, then, is a tangle of the corresponding set S of potential features? Formally, it will be a consistent specification of the elements of S, a choice of either $\vec{s}$ or $\overleftarrow{s}$ for every s such that each combination of up to three of these features can be found in at least one $v \in V$, and which avoids certain sets of features that we may choose and collect in a set $\mathcal{F}$. What, then, shall we choose as $\mathcal{F}$?

For a start, we will probably wish to include in $\mathcal{F}$, for some integer n, the set $\mathcal{F}_n$ defined in Section 2.3: this will ensure that each triple of features in a tangle can be found not just in one of our molecules but in at least n of them, which makes the tangle 'more typical' for V the larger we choose n.

In addition, we can use $\mathcal{F}$ to ensure that the collection of features encoded in a tangle matches the kind of DNA we wish to investigate. For example, we can force a tangle to choose at least one base for each position p by including in $\mathcal{F}$ the set $\{\,\overleftarrow{s}(p, A),\ \overleftarrow{s}(p, T),\ \overleftarrow{s}(p, G),\ \overleftarrow{s}(p, C)\,\}$ for every position p. If we wish to focus our investigation on organisms that have some known bases in certain positions, we can include in $\mathcal{F}$ some sets of features that specify different bases at such positions, thereby ruling out DNA from organisms we wish to ignore.

With such a set $\mathcal{F}$ in place, we may informally think of a tangle τ of S as assigning to each position a unique base so that this assignment is typical for some of the molecules encoded in V, to a degree determined by the agreement value n we require of our tangles. If τ has a guiding set $X \subseteq V$, then τ will be typical for the sequences in X also in the sense that in every position more than half of the sequences in X – or more than some higher proportion – have at that position the base specified by τ.

For every organism of which our sample contains sufficiently many molecules we can expect to have two tangles: one resembling each of the two strands of a DNA molecule from this organism. Each of these tangles will be a concrete hypothetical base sequence that is typical for some sequences in V (as above), and these two tangles will be sequences that are inverse to each other.[8] In the language of Chapter 2, these two tangles will be 'ideal' copies of (the relevant positions of) the two strands of a DNA molecule from this organism. Every actual strand from this organism in our sample may deviate slightly from this, due to some corruption of our data, but the 'ideal' base sequence will still be identified.[9]

If tangles of S correspond to individual organisms in this way, how are species or subspecies represented as tangles in this setup? The answer is simple: by tangles of subsets of S. For example, a given species would be identified by those DNA positions at which we expect the same specifications across all organisms of that species, but where other species might have different specifications. As with individual organisms and tangles of the entire S, a tangle of this subset of S will be an 'ideal' (partial) sequence of bases as they typically occur in DNA of that species. This tangle can be recognized from our sample of DNA even if no molecule in that sample has exactly these specifications in those positions (e.g., because our data is imperfect), since tangles are robust against changes of specifications $v(s)$ for just a few $v \in V$, or even against changes of some $v(s)$ for many v as long as not too many of them happen to the same s.

Now each tangle is a concrete and definitive specification of the positions it considers, and thus defines one concrete, if hypothetical, partial DNA molecule that is 'typical' for the species etc. which this tangle represents. It would therefore be interesting to compare these concrete representative sequences with those laid down in the data bases used for reference for the corresponding species.

As an application of the first main tangle theorem, we shall see in Chapter 12, Section 12.3, how the tangles at different levels interact. Tangles representing various subspecies of a given species, for example, will include the tangle representing that main species as a subset (of features). The theorem will show how all these tangles are organized in a tree-like structure. From this structure we can read off the phylogenetic tree of all the species, subspecies, organisms and so on that are represented in our DNA sample.

The second main tangle theorem will provide verifiable evidence proving, for example, that no single organism has more than a specifiable number of molecules in the sample – information that might be relevant in forensic contexts.

4.3 Drug development: chance discovery or focussed effort?

Innovation in pharmaceutical research sometimes benefits from chance discoveries: newly found or developed substances are observed to have properties that were, perhaps, not intended in the context in which these substances were studied, but which could be harnessed to beneficial effect in other contexts. In this section we look at how tangles can help with target-driven research by singling out groups of therapeutic targets that can be represented by a tangle and, in that way, be addressed together.

Suppose we are tasked with developing drugs against some types of harmful pathogens, so many that we cannot target them individually.

We might then look into ways of combining these pathogens into *clusters*: groups of similar pathogens that might be treated by the same drug, so that we only have to develop one drug for each group.

In order to know which pathogens we should put in the same cluster, we thus need a sensible notion for pathogens of being 'similar'. Now clearly, it may seem, two pathogens are similar if they share many features – features of whatever kind may be relevant for the development of suitable drugs.[10]

A moment's thought, though, shows that there is a snag here: we might well end up with clustering together pathogens that are indeed pairwise similar in that every two of them share many relevant features, but which features these are might depend on the pair of pathogens we are looking at. We might then be able to develop a common drug for every two of those pathogens. But there would be no guarantee that any of them would work for more than two, let alone all, pathogens in the same cluster.

We are thus thrown back to the beginning: in order to be able to, potentially, treat our many types of pathogens by just a few drugs, we need to cluster not the pathogens themselves but their features. More precisely, we would like to find a few groups of features that can each be targeted by one drug, and such that most of our pathogens have their own features represented sufficiently well by the features in one of these groups that the drug developed for that group works for this pathogen.

Applying the same standard clustering approach to features rather than pathogens we would now be looking at what makes two features 'similar' – so that they can be clustered together. These aspects of similarity, such as molecular composition or shape, may as such be independent of our pathogen sample. But in order to help with our task, our notion of similarity must also be borne out by our existing sample of pathogens in that features are regarded as similar only if they often occur together. If we ignore this, we shall end up with too many clusters and will need too many drugs.

Tangles offer a possible solution: they are groups of features that often occur together in the pathogens at hand. Although there is no guarantee that this happens always, we may expect that, conversely, most of our pathogens are represented reasonably well by a tangle, so that any drug developed for the features in that tangle also works for them.[11]

We shall follow up on this idea in Chapter 12, Section 12.4, when we have better formal grasp of what tangles are and how they may be used.

5

Examples from the social sciences

This chapter is written in the informal terminology of Chapter 2. We have a set V of objects or people, and a set $\vec{S}$ of possible features of these objects or people. These features come in *inverse* pairs $\{\vec{s}, \overleftarrow{s}\}$, which we call 'potential features' and denote by dropping those opposite arrows, in our case by s. The inverse features $\vec{s}$ and $\overleftarrow{s}$ are the two *specifications* of $s = \{\vec{s}, \overleftarrow{s}\}$. Every object $v \in V$ has exactly one of the features $\vec{s}, \overleftarrow{s}$ for every s; we denote the one it has by $v(s)$, and say that v *specifies* s as $v(s)$. We write S for the set of all s with $\vec{s} \in \vec{S}$.

When V is a set of people, S can be thought of as a questionnaire, with $\vec{s}$ and $\overleftarrow{s}$ as the two possible answers for a given $s \in S$, yes or no.[1]

A *specification* of S is a subset of $\vec{S}$ that contains either $\vec{s}$ or $\overleftarrow{s}$ for each $s \in S$, but never both. It is *consistent* if for every three of its choices, $\vec{r}, \vec{s}, \vec{t}$ say, there exists an object $v \in V$ that specifies r, s and t in exactly this way, as $v(r) = \vec{r}$ and $v(s) = \vec{s}$ and $v(t) = \vec{t}$. A consistent specification of S is a *tangle* of S if it has no subset that is an element of some set $\mathcal{F}$ of subsets of $\vec{S}$ that we may specify. When no such $\mathcal{F}$ is specified it will be $\mathcal{F} = \mathcal{F}_n$ for some n, with $n = 1$ as the default. When $\mathcal{F} = \mathcal{F}_1$, all consistent orientations of S are tangles.

The first section in this chapter differs a little from the others. Unlike those, it does not merely describe some concrete toy example of one particular tangle application, but it is somewhat broader in design. On the face of it, it shows how tangles can be used to detect mindsets in a population, collections of views that are often held together, from how these individuals answered a questionnaire. But the topic of this questionnaire does not matter for this section: our treatment will be generic for all questionnaire-based studies in the social sciences. In fact, it can serve as a template also for tangle applications outside the social sciences. We shall make this more concrete later; for the moment, let us just remember that Section 5.1 is more generic and therefore, perhaps, more important if less fun than the other sections in this chapter.

The remaining four sections then consist of some concrete tangle applications from different fields: psychology, analytic philosophy, political science, and education. These examples are also relevant more widely than just in their own context, but mainly by analogy, not because of any generality built into them.

5.1 Sociology: discovering mindsets and social groupings

Mindsets were the example of tangles we discussed in the introduction, where in Section 1.2 we considered a set of individuals polled with a political questionnaire S. The reason we called those tangles 'mindsets' was simply that S asked about opinions. In a different example, the tangles of S might be called something else – perhaps 'personality types' if S asks about traits of character, and so on. In what follows, these differences will be immaterial. So in order to have a vivid term for tangles of questionnaires, no matter what these are about, we shall always call such tangles *mindsets*, or *mindset tangles*.

Similarly, we may refer to any set S of potential features that have meaningful interpretations as a 'questionnaire', even if S consists of observations made by an external observer, of qualities as in our chair example, or of physical measurements as in Chapter 4. This is to distinguish such potential features from those considered in Section 1.3, which could be arbitrary partitions of V that had no meaningful interpretation.

Our mindset tangles in Section 1.2 were popularity-based in the sense of Section 1.1: they were guided by a set X of people that mostly 'subscribed to this mindset'. When we speak of mindset tangles from now on, we will not assume this: such tangles may, but need not, have a guiding set as defined in Section 2.5. The fact that mindsets can be identified from questionnaire data without having to identify like-minded groups of people first is a key property, and asset, of tangles: they are typical patterns of behaviour or views, not groups of people of similar behaviour or views.

This differs fundamentally from the classical approach that seeks to find social groups as clusters of people. Being defined directly in terms of the phenomena that define social groups, rather than indirectly in terms of the people that display them, tangles can bypass many of the usual problems that come with traditional clustering, such as the fact that an individual is likely to belong to more than one social group.[2] At the same time, unlike with the statistical methods found in the toolkit of sociologists, tangle mindsets can be found in data based on a general questionnaire that did not target, or test for, their existence. Tangles can help us discover mindsets, not just confirm or disprove our conjectures about their existence.

We shall mostly assume that S consists of yes/no questions only. If it does not, we can simply translate its answers into answers of an equivalent imaginary questionnaire of yes/no questions, which we then use as the basis for computing our tangles.[3] It is also possible to allow questions with a range of possible answers directly, see Section 7.6. But most of the theory and potential applications of tangles discussed in this book can be explained in the more intuitive scenario of yes/no questions, and so this is what we shall do.

Specifying the collection $\mathcal{F}$ of 'forbidden' sets of features is our main tool for determining how broad or refined will be the mindsets that show up as tangles. If we forbid no feature sets, i.e., leave $\mathcal{F}$ empty, then every specification of S returned by one of the people polled will determine a tangle, because it is consistent.[4]

Such tangles would be too specific, or 'focussed', to be helpful. If we do not wish to influence which tangles are returned by our algorithms other than by determining how broad the mindsets should be which they define, we can choose $\mathcal{F}$ as the set $\mathcal{F}_n$ of all triples $\{\vec{r}, \vec{s}, \vec{t}\}$ shared by fewer than n people polled, to ensure that every three answers specified by a tangle were given by at least n people in V. We might even experiment with this agreement parameter n at the computer and see how it influences the number, and degree of focus, of the tangles we obtain.

But we can also decide to add some sets of features to $\mathcal{F}$ that we think of as inconsistent because of what they mean. What these features are will be up to us: we might add sets of features that are intuitively inconsistent, or sets of features that will occur together as answers on the same questionnaire only if someone tampered with it and tried to influence the survey.

We might even use $\mathcal{F}$ to deliberately exclude some types of mindsets from showing up as tangles, e.g., mindsets that simply do not interest us. In order to exclude such mindsets we simply add to $\mathcal{F}$ the feature sets that make them uninteresting: then no specifications of S that include these feature sets will show up as tangles.[5] Of course, subsets we forbid for this reason must be specific enough that they occur *only* in mindsets that are not of interest to us.

To summarize, let us emphasize again what makes the tangles approach to the study of mindsets, social groupings, or types of character etc. different from traditional approaches in sociology that are less structural and more statistics-based. It is that we can mechanically find these mindsets etc. from observational data, without any prior intuitive hypothesis of what they may be, and that we can identify them without having to group the people we observed accordingly. This should make it possible to discover hitherto unknown mindsets in sufficiently rich data.

In Chapter 13 we shall apply the tangle theory we shall have met by then to find structure also between those mindsets: how they can be distinguished, or how one refines another. Our algorithms described in Chapter 11 can do this mechanically too: they can find both hitherto unknown mindsets and the structure between them without any need to guess those mindsets first, or to interpret the tangles found before we wish to do so.

5.2 Psychology: understanding the unfamiliar

This example is, essentially, a special case of the mindset example from Secion 5.1. But the special context may lend it additional relevance.

While it is interesting to search for combinations of, say, political views that constitute hitherto unknown mindsets that can have an impact on political developments, it is not only interesting but crucial to try to *understand* minds that work in ways very different from our own. This is particularly relevant in doctor–patient relationships where the doctor seeks to offer an individual with such a different mind a bridge to society, or even just to their particular environment: to enable them to understand the people around them, and to help those people to understand them.

Tangles can already help bridge this gap at the most fundamental level, the level of notions into which we organize our perceptions. We all have a notion, for example, of 'threat'. But some patients' notion of threat may be different from ours: they may be scared by things we would not see as threatening. And what these are may well come in types: typical combinations of perceptions of everyday phenomena which, for people with a certain psychiatric condition, instil fear, and thereby shape their notion of 'threat', one that may well differ from ours.

Tangles are designed to identify such types of perceptions, some that combine to notions common to patients with some particular psychiatric condition but which may be unfamiliar to us. The set $\vec{S}$ would consist of various possible perceptions that experience has shown are relevant for the symptom – in our example, fear – that we are trying to explain; the set V would consist of patients whose actual perceptions regarding the potential perceptions s in S are on record; and a tangle would identify the combinations of these that are typical for psychiatric patients in the sense of Section 2.3.

Such combinations may then be interpretable as notions – such as of 'threat' – that are shared by patients with similar psychiatric conditions but may be unfamiliar to us. Indeed, once such tangles have been identified, we can train our intuition on them in an effort to build such an interpretation as a 'notion' in our own minds, and thereby to *understand* our patients rather than just collect lists of unrelated symptoms.

We shall get back to how tangles can help identify meaning in Section 5.3. The emphasis there, however, will be on the diversity of the things we bundle together to form a notion, and possibly refer to them all by some common word. So the set V whose typical features we examine there will be a set of things; not, as in our example here, a set of people.

At a higher level, tangles can help us identify entire psychological syndromes, and maybe discover hitherto unknown syndromes. Indeed, psychological syndromes appear to be exactly the kind of thing that tangles model: collections of features that often occur together.

This would be true for any medical condition, or even for mechanical 'conditions' that lead to the failure of a machine. Indeed, as we saw in Section 4.1, tangles can be applied to build expert systems to assist in diagnostic processes. But what makes psychiatric conditions even more amenable to the use of tangles is that the symptoms of which they are combinations are so much harder to quantify. Tangles come into their own particularly with input data that cannot be expected to be reliably precise.

So how would we formalize the search for hitherto unknown psychological syndromes? Our ground set V would again be a large pool of patients in some database. But $\vec{S}$ would now be a set of possible symptoms, not – as in our earlier example – potential indications or expressions of some fixed symptom such as fear. Tangles will be collections of symptoms that typically occur together: medical, or psychological, conditions, traits, or illnesses.

We shall see in Chapter 13 that the set T of tangle-distinguishing features returned by the first of the two main tangle theorems will consist of 'critical' symptoms or combinations of symptoms that can be tested on a patient in the process of diagnosis. Every consistent specification of T defines a unique tangle: a unique condition that has all the symptoms in this specification of T. In most cases, this will be the correct diagnosis.

Note, however, that our emphasis here is on *finding* psychiatric conditions: our aim is to discover combinations of symptoms that constitute an illness. It is not on diagnosing a given patient with one of these conditions (although checking the symptoms in T can play an important part in this), but on establishing what are the potential conditions to look for in the diagnosis. A lot has happened in psychology here in recent years, and conditions are now recognized as illnesses that were not even thought of as typical combinations of symptoms – tangles – not so long ago.

5.3 Analytic philosophy: quantifying family resemblances

What is a chair? What does the word 'chair' mean? What *is* meaning?

Dictionaries are meant to tell us the meaning of a given word, such as 'chair'. If we looked up 'chair' in a dictionary a hundred years ago, we would likely be given a list of properties that a thing should or should not have in order to warrant being called a chair. In other words, the dictionary would give us a list of predicates such that, ideally, a given thing would satisfy all these predicates if and only if it was a chair.[6] Even today, this paradigm forms the basis of most approaches in computer science to capture natural languages, such as in formal concept analysis [22].

But we all know that this does not work in practice. The way we use words, and hence what they refer to, is not set in stone: meaning is constituted in a social process. Our use of a word is often determined by intuitively-felt analogy, which depends both on us, the individual, and on the context. This is true for single words, and more so for entire phrases, not to mention non-verbal communication such as body language.

Apart from its impracticality, the traditional dictionary approach to meaning is also more fundamentally flawed. One objection is that it begs the question: being a chair is a predicate as much as those expected to explain its meaning. In order to help us understand the meaning of 'chair', the meaning of those other predicates should already be known to us. Those that are not should be defined by their own list of even more basic predicates, and so on. What if this process never ends? Indeed there is no reason it should. As anyone knows who has tried to explain something complex to someone entirely unfamiliar with the context, it can be quite difficult to find sequences of explanatory predicates that are *not* circular.

Even if the risk of circularity in dictionary-like explanations does not bother us too much – after all, it might happen that we do know all the predicates that our dictionary employs to tell us what a chair is – there is another objection to this approach, equally fundamental: a list of predicates to explain 'chair' as envisaged might simply not exist.

One reason for this could be that our language is not rich enough. Since all chairs, but only chairs, should satisfy all the predicates on the desired list, the list will have to contain, for every thing that is not a chair, some predicate that all chairs satisfy but this thing does not satisfy. Our language might not contain enough predicates for this to be possible.

But there is another, more fundamental, problem. The entire approach is based on the assumption that every predicate in our language corresponds to a set of things: those that satisfy the predicate. The set of chairs would then be the intersection of the sets of things that satisfy the various predicates in our list. But that is far from true. We can probably make a thing that would divide opinion about whether it is a chair or not. Similarly, predicates such as 'sturdy' or 'comfortable', which might well come up in a definition of 'chair', are clearly not just tags for sets of things that are well defined in the sense that we would all agree, for every object shown to us, whether or not it lies in that set. Thus, words simply do not correspond to well-delineated sets of things, which they 'merely name'.

Wittgenstein [35] recognized this, and said that the things we would use a certain word for are more like the members of a family: some share one feature, others share another, and there may not be a clearly delineated set of precisely the things that are referred to by that word.

Tangles formalize exactly this.[7] Let τ be the set of predicates that our dictionary claims define the notion of 'chair'. For every $\vec{s} \in \tau$ let $\overleftarrow{s}$ denote its negation, write $s := \{\vec{s}, \overleftarrow{s}\}$, and let $S := \{s \mid \vec{s} \in \tau\}$. Then τ is likely to be a tangle of S, even one of the kind we called 'popularity-based' in Sections 1.1 and 2.5. For, unlike what we expect of a traditional dictionary definition, this no longer requires that every chair satisfies all the predicates in τ, and that no other thing does. There need not even exist a single chair that satisfies them all: it will be enough that each predicate in τ applies to something like eighty percent of 'all chairs'. And because this requirement is so robust against disagreement in a few cases, the 'set of all chairs' need not be properly delineated: we need not be able to decide for every thing x and every predicate $\vec{s}$ in τ whether $\vec{s}$ applies to x. If we cannot agree, we simply pick an answer. Then perhaps those eighty percent come down to seventy – which is still good enough for τ to be a tangle.

Note that having the notion of a tangle does not enable us to decide any better than before whether a given object is or is not a chair. The whole point is that this is too much to ask: we have a working notion of 'chair' without being able to identify all individual chairs without ambiguity or doubt. Wittgenstein intends his analogy between meaning and family resemblances to express just this: he argues that our inability to come up with a dictionary-type definition for a word in no way implies that our notions are meaningless. On the contrary: meaning can exist without an extensional equivalent, a well-defined set of things. Tangles, now, go a step further: unlike family resemblances they are precisely and formally defined. With this precise definition they can capture things like the meaning of 'chair' that are intrinsically vague, or 'fuzzy', at the extensional level, that of sets of things.

> Tangles offer a precise way to define extensionally
> imprecise meanings of words.

Indeed, more than one way: adjusting, for example, their agreement parameter n,[8] we can obtain many tangles of our set S of predicates each offering a definition of 'chair'. This, however, is again in complete correspondence with how we use words: we can use the same word loosely today, and more narrowly tomorrow.

Although every specific tangle in our context here is a concrete list of predicates, the notion of 'tangle' as such is purely formal and makes no reference to such predicates: a *tangle* is a specification of S that satisfies certain formal requirements: consistency and, perhaps, not containing certain small sets of features (which in our context are predicates). These requirements ensure that tangles define 'good' notions rather than 'bad' ones. We shall come back to this, and more generally to the problem of how to formalize 'meaning', in Section 6.5, where we look at it from a more practical computer or AI perspective.

In Chapter 7, Section 7.4, and in Chapter 9, we shall see how we can capture the meanings of different words by tangles of some common set of features – not a set chosen specifically, as in our example above, for the word whose meaning we intend to describe. That served its purpose here, which was to show how tangles can quantify Wittgenstein's family resemblances. But in less philosophical contexts it will be good to consider, and compare, different tangles of a single set of features.

5.4 Politics and society: appointing representative bodies

Finding the tangles of a questionnaire S, a collection of questions answered by some set V of individuals, is not primarily aimed at grouping V into sets of like-minded people, but at finding their prevalent mindsets as such: ways of answering S that are typical for the people in V. Once we have determined these prevalent mindsets, however, we may well be interested in finding individuals in V whose views reflect those mindsets best. For example, such individuals might represent V on a body that makes decisions related to topics addressed in S.

Following this idea, we might appoint the delegates to some decision-making body D by first conducting a survey S among the electorate V for D, then determining its mindset tangles, and finally appointing to D one representative for each tangle τ of S: for example, the person $v \in V$ for whom $|\{\, s \in S : v(s) = \tau(s)\}|$ is maximum amongst all the people in V. We shall discuss more refined versions of this idea in Chapter 13.

This process of constituting a representative body contrasts with the usual democratic process of electing delegates by majority vote. The traditional first-past-the-post system with constituencies, for example, can generate parliaments with large majorities even when, in each constituency, the majority of the successful candidate was small, but the population is homogeneous enough across constituencies that in most of these the successful candidate is of the same political hue.

In systems with proportional representation this is avoided, but such systems require the prior establishment of political parties. Even when these exist, they may have developed historically in contexts that are no longer relevant today. Finding the tangles of a political questionnaire is like *finding* the political parties that ought to exist today for the elections at hand. People $v \in V$ representing these virtual 'parties' as indicated can then be delegated to our decision-making body straight away, without any further election, since the system ensures that they represent those virtual parties best, at least in terms of their views.

If we wish to appoint more delegates for mindsets with a larger following, we could elect representatives of such a virtual party, a tangle τ of S, by a standard vote conducted only among its 'members', the people whose views are closest to τ amongst all the tangles of S. Alternatively we could refine τ into smaller tangles (see Sections 7.4 and 13.3), so that all tangles end up with roughly the same amount of support and could thus be represented by one delegate per tangle.

This approach may be even more relevant outside politics, where there are many situations on a smaller scale in which we seek to appoint a decision-making body. Think of the governors for a school, or a steering committee for a choir. At such a smaller scale there will be no constituencies. And there may be no established parties relevant to the brief of that body, or even formal lists of like-minded candidates that have banded together in some formal way preceding the election. But appointing for every tangle τ of S some members of V whose views on S are closest to that tangle would produce a committee likely to represent the views held in V well.

5.5 Education: combining teaching techniques into methods

This application starts from the assumption that different teaching techniques work well or less well with different students: that any given technique is not necessarily better or worse than an alternative for all students at once, but that each may work better for different sets of students.

Ideally, then, each student should be taught by precisely the set of techniques that happen to work best for him or her. Of course, this is impractical: there are so many possible combinations of techniques that most students would end up sitting in a class by themselves. But our aim could be to group techniques into, say, four categories we might refer to as *methods*, one to be used in each of four classes held in parallel, so that each student can attend the class whose method suits them best.

The question then is: how do we divide the various techniques into such groups? Just to illustrate the problem, consider a very simple example. Suppose first that some students benefit most from supervised self-study while others are best served by a lecture followed by discussion in class. Suppose further that some students understand a grammatical rule best by first seeing motivating examples that prepare their intuition, while others prefer to see the rule stated clearly to begin with and examples only afterwards. So there are four possible combinations of techniques, but maybe we can only have two classes. So we might try to group our techniques into pairs.

How shall we select the pairs? Shall we have one class whose teacher lectures and motivates rules by examples first, or shall we group the lecture-plus-discussion class with the technique of introducing rules before examples?

This is where tangles come in. Think of V as a large set of students whose performance has been evaluated for our study, and of $\vec{S}$ as a set of teaching techniques. Every tangle of S will be a particular combination of techniques that work well together for some students. If the tangle is guided by a set $X \subseteq V$ (see Section 2.5) then this X is a group of students that would benefit particularly well from the combination of techniques specified by this tangle: for every technique in this tangle, a majority of the students in X prefer that technique to its converse. Thus, X would be an ideal population for the class in which the techniques from this tangle are used, while a set Y guiding another tangle would consist of students best served by the teaching techniques specified by that tangle.

Once more, speaking about tangles in terms of guiding sets helps to visualize their benefits, but it is not crucial: the benefit arises from finding the tangles as such, and setting up those four or five classes accordingly. We then *know* that this serves our students best collectively – and we can happily leave the choice of class to them.

Of course, although we referred to our tangle-based groups of techniques as 'methods', they need not be related in any deeper way, such as follow some particular philosophy or didactic concept. They are simply evidence-based combinations of techniques that happen to work particularly well for some students, not a substitute for proper pedagogy.

When we return to this example in Chapter 13, we shall have to address the question of how to set our tangle parameters in such a way that, even if the set $\vec{S}$ of potential teaching techniques is large, we still end up with a desired number of tangles – four in our case – that may be dictated by the environment in which the teaching is taking place. We shall also see how, if we do not wish to let the students choose their class themselves, the first of our two tangle theorems can help us devise an entry test that assigns students to the class that benefits them best.

6

Examples from data science

In this chapter we pick up the story from Section 1.3, using also terminology from in Chapter 2. We have a dataset V of 'points', and a set S of partitions $\{A, B\}$ of V into two *sides* A and B, which we assume to be non-empty.

We usually denote the sides of a partition s as $\vec{s}$ and $\overleftarrow{s}$, so that $s = \{A, B\} = \{\vec{s}, \overleftarrow{s}\}$.[1] We think of the arrows as 'orienting' s towards the sides they denote, and also call these the two *orientations* of s. We write $\vec{S}$ for the set of all orientations, or sides, of elements of S. Given any $s \in S$, every $v \in V$ lies in exactly one of the sets $\vec{s}, \overleftarrow{s}$; we denote this side of s as $v(s)$.

An *orientation* of S is a choice of one side of s for every $s \in S$: a subset of $\vec{S}$ that contains exactly one of $\vec{s}, \overleftarrow{s}$ for every $s \in S$. An orientation τ of S is *consistent* if every three of its elements have a non-empty intersection. For example, every orientation of S of the form $v(S) := \{v(s) \mid s \in S\}$ for some fixed $v \in V$ is consistent.

A consistent orientation of S is a *tangle* if it has no subset that is an element of some set $\mathcal{F}$ of subsets of $\vec{S}$ that we may specify. When no such $\mathcal{F}$ is specified it will be $\mathcal{F} = \mathcal{F}_n$ for some n, with $n = 1$ as the default. When $\mathcal{F} = \mathcal{F}_1$, then all consistent orientations of S are tangles.

We begin in Section 6.1 by applying tangles to generic clustering: without making any application-specific assumptions about S, we look at ways of turning tangles of S into clusters in V, assuming only that the partitions in S are indeed at 'bottlenecks' of V.[2]

In Section 6.2 we apply this more specifically to clusters in images: we look at ways in which homogeneous regions of an image can be captured by tangles. At this basic level, the only difference to the generic clustering of Section 6.1 will be that we can use the fact that V contains an image to come up with certain natural ways of choosing S. We shall discuss those, and their influence on the tangles we obtain, in Section 6.2. In Chapter 14 we shall return to this topic and discuss how the tangle theory we have met by then can help us put those tangle-induced regions back together to form a computer's memory of the objects shown in an image, leading to new approaches to image recognition and compression.

In Section 6.3 we apply tangles to purchasing data of an online shop, with V the set of customers and S the set of items on offer. Tangles of S will be types of items often bought together. Different tangles arise, for example, from different motivations in shopping: to save money, to favour ecological products, or to look for exciting new things. The online shop example is interesting also for another reason. It illustrates the use of what we shall come to call 'duality of feature systems': in the interaction between customers and products defined by the purchases on record we can see not only tangles of S, reflecting types of products, but also tangles of V reflecting something like customer types, or typical purchasing behaviour. When we describe this duality mathematically in Chapter 7, we shall refer back to this example.

In Section 6.4 we look, quite literally, at 'tangles of words': how tangles can help us identify topics, genres, or authors of texts. Unlike the use of tangles in image analysis, this will not be an application of tangle clustering, quite the opposite. In the terminology of Sections 1.1 and 2.5, text tangles will be 'consistency-based' rather than 'popularity-based', even more than our mindset tangles in Section 5.1. Text tangles, therefore, offer a genuinely new approach to the classification and identification of texts, one different from approaches based on clustering.

In Section 6.5, finally, we look how tangles might help with semantic text analysis. We shall pick up the thread from Section 5.3, where we saw how tangles can be used to formalize Wittgenstein's approach to our notion of 'meaning' by his now famous 'family resemblances'. We shall refine and develop this further in Chapter 14, Sections 14.9 and 14.10, by applying the tangle theory from Chapter 8 to construct an interactive thesaurus.

6.1 Indirect clustering by separation: hard, soft or fuzzy

In Section 1.3 we motivated the idea behind tangles by showing that point clusters, whatever their precise definition may be, orient certain 'bottleneck' partitions of V in a consistent way, namely, towards the cluster. The underlying assumption was that, given any such cluster $X \subseteq V$ and any such 'bottleneck' partition s of V, most of X would lie one side of s, say on the side $\vec{s}$, in which case X would orient s as $\vec{s}$.

The idea, then, was that these 'consistent' orientations of all the 'bottleneck' partitions of V, which we would call a 'tangle', might constitute the essence of the cluster X that defined them: that this 'tangle' would be invariant under changing X a little bit, or changing the definition of 'cluster' a little bit, or adding some noise to our data, but at the same time retain enough information to enable us to reconstruct from it the fact that there is a cluster in V somewhere near X.

In order to get this indirect approach to clustering to work, however, we had to come up with definitions for all the ingredients used in the approach that were independent of any pre-conceived notion of a cluster. We thus had to define bottleneck partitions, consistency of orientations, and ultimately tangles, all independently of any earlier notion of cluster, but in a way compatible with those cluster-induced meanings of these terms that motivated them in Section 1.3.

So, now that we do have a formal definition of tangles, as repeated in the introductory text for this chapter, does it live up to this promise?

Well, if we assume for the moment that our set S of partitions of V does consist of 'bottleneck' partitions, it certainly lives up to the requirement of being formally independent of our guiding intuition: our definition of consistency of orientations of S makes no reference to any pre-conceived notion of a cluster, and neither does our notion of tangle, which simply allows us to narrow down the class of consistent orientations of S further if we wish. Moreover, as pointed out in Section 2.5, any set $X \subseteq V$ that is sufficiently cluster-like to ensure that more than two thirds of it lie on one side of each partition in S, our set of 'bottleneck' partitions, defines such a tangle.

Our aim in this section, then, is to address the converse question: can we use our tangles to find clusters in V – perhaps even associate with every tangle of S a cluster X in V that defines this tangle as above?

The other remaining question, that of how to define 'bottleneck' partitions without reference to any pre-conceived notion of a cluster, will have an entire chapter devoted to it later, Chapter 9. For now, let us simply think of the elements of our given set S as the bottleneck partitions of V, whatever these may be in a particular context. However, we shall continue to appeal to the intuitive picture of Figure 1.3, where the bottleneck partitions include all those that divide the point set at the 'handle', an elongated isthmus between one cluster on the left and three clusters on the right.[3]

Let us now address the topic of this section: the question of how we might define clusters in V from tangles of S that we have computed. We shall look at three types of clusters, which in data science would be called 'hard', 'soft', and perhaps 'fuzzy' clusters.

Hard clusters are disjoint subsets of V whose elements are, somehow, close to each other. In order to find a hard cluster associated with a given tangle τ, we shall ask separately for every $v \in V$ with which tangle, if any, it might best be associated with. The hope, then, is that points associated with the same tangle are close to each other in some sense.

Whether or not that will be the case has to be tested by experiment: our definition of these point sets associated with a tangle will be formal and precise, but since the notion of being 'close to each other' is not formal, there will be no formal way of proving that such tangle-induced subsets of V are indeed (hard) clusters. But such experiments may include what physicists would call thought-experiments: constructions of examples that highlight different possible aspects of our candidates for tangle-based clusters. We shall see such theoretical examples below.

Our soft clusters will be defined similarly. As with hard clusters, we consider each $v \in V$ separately and look for tangles with which to associate it. The difference now is that v does not have to choose just one such tangle: it may opt to be associated with several tangles to varying degrees. Our tangles can then be thought of as 'clusters' in which each $v \in V$ has a certain degree of membership.

Fuzzy clusters, finally, are again concrete subsets of V. But they are not defined in terms of properties shared by their individual elements, but directly by a property of subsets of V. In our case these will be properties of subsets of V that relate them to a tangle. Whereas any given $v \in V$ can, by definition, lie in at most one hard cluster, it may well lie in several fuzzy clusters: different point sets associated with the same tangle, or even with different tangles. But whenever it lies in such a set it does so properly, not just 'by degree' as in the case of soft clusters.

Let us start with two such tangle-inspired ideas for fuzzy clusters. In Section 2.5 we met two properties of subsets of V that associate these with a given tangle of S: the properties of *guiding* a tangle, and that of of *witnessing* one. Are such subsets of V intuitively cluster-like?

Since the entire point set V witnesses every tangle, tangle-witnessing sets as such will not in general be cluster-like. Tangle-guiding sets, as such, are also not in general cluster-like. In Section 2.4 we already met an extreme example for both these facts, the 'black hole' tangles. These are centred on a 'void' rather than on a cluster, yet are guided and witnessed by all the points floating around that void, no matter how far away. In fact, any subset of V that witnesses our black hole tangle from Figure 2.1 meets at least four of its five natural clusters, and any subset of V that guides it meets at least three of them. Hence even minimal witnessing or guiding sets are not always cluster-like.

Another caveat about considering tangle-guiding sets as candidates for natural clusters is that any given $v \in V$ will typically lie in *some* guiding set for *every* tangle τ that has a guiding set at all, since adding v to such a set X will not invalidate it as a guide for τ: it will only reduce its reliability a little bit if $v(s) \neq \tau(s)$ for some s on which X already agrees least with τ. Looking for natural clusters among guiding sets, therefore, will not immediately give us a way of associating given elements of V with any particular such cluster.

However, subsets of V that simultaneously guide and witness a given tangle will remain at the top of our list as suspects for natural fuzzy clusters. Recall that our informal notion of 'popularity-based' tangles in Sections 1.1 and 2.5 was inspired by such sets, sets which guided a tangle with a reliability of more than two thirds and therefore also witnessed that tangle. We shall discuss guiding and witnessing sets of tangles as candidates for fuzzy clusters in more detail in Chapter 14. In particular, we shall identify some additional conditions which cluster-like sets that witness and guide a tangle tend to satisfy, but which the set of peripheral points in the black hole tangle does not satisfy.

Next, let us look at possible ways of associating with a given tangle some natural hard cluster in V, a set of points which, somehow, lie close together. More sophisticated ways of using tangles for hard clustering, one based on Theorems 1 and 2 from Chapter 8, and another based on 'advanced feature' tangles, will be described in Chapter 14, Sections 14.1 and 14.5.

In order for our clustering model to capture reality, not every element of V should have to lie in a hard cluster. Indeed, in contrast to some of the best-known existing clustering algorithms, it should be possible that

V contains no cluster at all if that fits our data best. This will certainly be possible for the tangle-induced hard clusters introduced below and discussed further in Chapter 14, if only because it can happen that S has no tangles. Theorem 3 in Chapter 8 will then describe the structure of our data in a way that allows us to *prove* that there are no tangles, and hence no tangle-induced clusters.

If our set S of partitions of V does have tangles, then these can be naturally compared with individual points in V.[4] Indeed, every $v \in V$ orients every $s \in S$ towards the side $v(s)$ that contains v, and every tangle τ of S likewise orients every $s \in S$ towards one of its sides, $\tau(s)$. We can therefore measure directly the extent to which a given $v \in V$ agrees with a given tangle τ, e.g. by counting the partitions $s \in S$ such that $v(s) = \tau(s)$. We might then associate each $v \in V$ with the tangle that fits it best.

Formally, we might define these point-tangle similarities as

$$\sigma(v, \tau) := |\{\, s \in S : v(s) = \tau(s) \,\}|,$$

or as

$$\sigma(v, \tau) := |\{\, s \in S : v(s) = \tau(s) = \vec{s} \,\}|$$

if the partitions s have default orientations $\vec{s}$ which, viewed as features, are more significant than their inverses $\overleftarrow{s}$ (as in Section 6.3 below). In either case let us associate with each $v \in V$ the tangle τ of S for which $\sigma(v, \tau)$ is maximum, and call it τ_v^+.[5]

Given any tangle τ of S we might then, conversely, look at the set

$$X_\tau := \{\, v \in V \mid \tau = \tau_v^+ \,\}$$

of points associated with τ in this way.

Natural clusters in V, subsets which define a tangle τ as in Section 1.3, will largely lie in the set X_τ for this τ, since their points v will not normally fit any other tangle better under our similarity function σ. But these X_τ will often be larger than those clusters. This is because, globally, the sets X_τ make up all of V between them (as long as S has at least one tangle), while not every $v \in V$ will lie in a natural cluster.

To make up for this, let us trim our sets X_τ a bit to increase their chance of being natural clusters: for each τ let us include in X_τ only elements v of V for which $\sigma(v, \tau)$ exceeds a certain threshold. If there is no such v then $X_\tau = \emptyset$, and we abort our attempt to find a natural cluster associated with this tangle τ. Conversely, those v for which $\sigma(v, \tau_v^+)$ falls below our threshold will not lie in any X_τ. This corresponds well to the reality we are trying to model, where not all points lie in clusters. Our threshold can be adjusted interactively when we compute our tangles, to produce whatever seems like the most cluster-like sets X_τ.

With such thresholds in place, can we identify any general conditions under which the non-empty sets X_τ are likely to be natural clusters? Picking up the thread from our discussion of fuzzy clusters, will it help to know that X_τ guides or witnesses τ?

We shall illustrate this question by two examples: one where X_τ witnesses τ but does not guide it, and one where X_τ guides τ but does not witness it. In neither case is X_τ a natural cluster. So these examples show that imposing just one of those two conditions on X_τ will not ensure it is a natural cluster. We shall then look into imposing both.

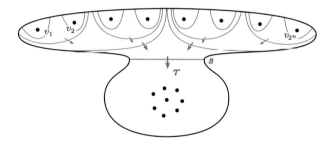

FIGURE 6.1. $X_\tau = V$ witnesses but does not guide τ. There are n points in the bottom pouch and 2^n in the top row.

Figure 6.1 shows an example where S has only one tangle τ.[6] For this tangle τ we have $X_\tau = V$, even with a threshold as high as about $|S| - \log |S|$, since $\sigma(v, \tau) = |S|$ for all v in the bottom pouch and $\sigma(v, \tau) = |S| - \log(|S| + 1)$ for all v in the top row. As it does for every tangle, V witnesses τ. But it does not guide τ, since $v(s) \neq \tau(s)$ for most $v \in V$ for the partition s indicated in the figure. However, $X_\tau = V$ is not a natural cluster in V in terms of S.[7]

In Figure 6.2 the partitions shown have four tangles for $n = 2$. The central one, τ, is guided by its set $X_\tau = \{v_1, v_2, v_3\}$, but not witnessed by it since $\vec{s_1} \cap \vec{s_2} \cap \vec{s_3} = V \smallsetminus X_\tau$. As before, X_τ is not a natural cluster.[8]

From now on, let us focus our hunt for hard clusters associated with a tangle τ on sets X_τ that guide *and* witness τ. Let us combine this with our earlier observation that it may be good to limit X_τ to those of its points v for which $\sigma(v, \tau)$, the number of elements s of S on which v agrees with τ, exceeds a certain threshold we may impose.

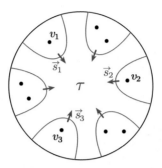

FIGURE 6.2. $X_\tau = \{v_1, v_2, v_3\}$ guides but does not witness τ.

Dually, we might require that, for every $s \in S$, the number of points in X_τ that agree with τ on s should exceed a certain threshold. If we phrase this threshold as a proportion of $|X_\tau|$, this is just our old notion of reliability of tangle-guiding sets from Section 2.5: recall that $X \subseteq V$ guides τ (if it does) with *reliability* p if p is maximum such that, for every $s \in S$, at least $p|X|$ of the points in X agree with τ on s.

Summing up, we have focussed our hunt for hard clusters related to a tangle τ on sets of the form X_τ that both guide and witness τ, just as we did earlier in our hunt for fuzzy tangles. We can try to increase the chance that these X_τ are cluster-like by imposing minimum thresholds: either for every $v \in X_\tau$ on the number of $s \in S$ with $v(s) = \tau(s)$, or for every $s \in S$ on the number of $v \in X_\tau$ with $v(s) = \tau(s)$, or both.

In Chapter 14 we shall meet another way to detect tangle-related clusters, one that depends on our tangle theorems from Chapter 8.

Clusters that are fuzzy in the *informal* sense that they cannot be clearly delineated as point sets in V cannot, by definition, be captured as hard clusters. This has given rise to the notion of *soft clustering* in data science, whose idea is as follows.

Hard clusters X, being subsets of V, have the property that every $v \in V$ either belongs to X or not, and usually we demand that no $v \in V$ belong to more than one such cluster X. In soft clustering we allow every $v \in V$ to belong to more than one cluster 'to some degree'. For example, the two red points in Figure 1.1 might find it difficult to make up their minds as to whether they belong to the top-left cluster or the middle one, and choose to belong to both of them to equal degree, or perhaps with a preference ratio of 55% to 45% for one cluster over the other.

We can think of the tangles of S as such soft clusters by giving every $v \in V$ a total weight of 1, say, to distribute over the various tangles of S – for example, in a way that reflects its similarities $\sigma(v, \tau)$ with the various tangles τ. Alternatively, we might associate every $v \in V$ with each tangle τ directly with a *confidence* of $\sigma(v, \tau)$, without requiring it to distribute some constant amount of weight amongst the tangles.

Figure 6.3 shows, for each of eight tangles found in partitions of a set of points in the plane, how much each of the points agrees with that tangle. The points that agree most with a given tangle form a soft cluster.

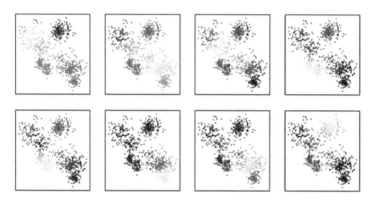

FIGURE 6.3.   Soft clusters: the more a point agrees with the given tangle, the brighter it is shown.

6.2 Tangles of pixels: clusters in images

In this section we look at how tangles can be used to identify regions in an image: a face on a picture, say, and an eye within that face. In Chapter 14, Section 14.6, we shall discuss how the mathematics of tangles, such as the tangle theorems to be presented in Chapter 8, can take this approach further and show us how those regions relate to each other to form a larger image.

Let us think of an image as a set V of pixels arranged in a grid-like pattern, and call V the *canvas*. What exactly these pixels are will depend on the intended application: they might be simply black or white, which we might encode as 1 or 0, or they might be 'vectors', tuples, of many parameters describing various aspects of their colour, brightness and so on. All that matters to us is that we can tell how similar two pixels are: that we have a *similarity function* $\sigma : (p, q) \mapsto \mathbb{R}$ defined on the pairs of pixels. We assume that σ is symmetric, i.e., that $\sigma(p, q) = \sigma(q, p)$ for all

2-element sets $\{p, q\} \subseteq V$, and that σ takes higher values when p and q are similar but lower values when they are not.

The visually identifiable regions in an image can be thought of as examples of clusters, sets of similar pixels. But not any such sets: the elements of the set of all red pixels are always similar (in this case, red), but if they are scattered around over the entire canvas we will not see them as a region. Thus, regions are not merely clusters in terms of σ; proximity of pixels also plays a role.

If our aim is to identify visible regions on our canvas by tangles, we can take care of this problem when we choose the set S of partitions of V which our tangles will have to orient. In the example, we would try to ensure that subsets of V that consist of points randomly scattered around the canvas are not selected as elements of $\vec{S}$: that the partition of V into the red points and all the others is not in S, unless there is a red blob on the canvas that deserves being called a region.

Rather, our set S of partitions of V should consist of what we shall call *cuts*: partitions of our canvas into two sets separated by a single line L that either runs between two points on the edge of the canvas or is closed, like a deformed circle.

In order for a tangle of S to capture a visible region, S should not contain all cuts. It should contain only cuts that look natural on the canvas: cuts whose separating line L does not have too many twists and turns, and such that pairs of pixels that are close but divided by L tend to be dissimilar. In the example of just black and white pixels we might require that at least about half, say, of the pairs of pixels that are adjacent and whose common boundary lies on L have different colour.

Figure 6.4 shows four lines L defining such a cut. The vertical black line on the left divides too many similar pixels adjacent to the line from opposite sides for us pick the corresponding partition of V for our set S of suitable cuts. The blue line is a little better, and might make it into S if we are generous: its short vertical segment has very different pixels to the left and right, while its middle segment following Mona Lisa's chin has a good number of light-dark pairs of near pixels above and below the line (although these are not directly adjacent). The green line is even better: on the left, most pairs of near pixels are very different, while on the right there are both similar and dissimilar close pairs. The red line around the eye, finally, would meet our highest standards for a good cut to be adopted into S: inside the red oval all pixels are dark, while outside it most pixels in the vicinity of the line are light.

FIGURE 6.4.   Lines of varying clarity cutting the canvas.

When we generate our set S of cuts whose tangles we plan to compute, it is important that we take not just *some* of the cuts of at least our desired degree of clarity, but all of them. For example, our benchmark for the most basic cuts to be put in S should be low enough that cutting through a large area R of random pixels counts as a good cut, one to be put in S: such an area is not one we wish to detect as a 'region', and so we do not want it to guide a tangle (see Section 6.1). We can ensure this by putting in S enough cuts right through R: then only tangles of agreement lower than we require, such as the focussed tangles, can orient them all consistently; compare Section 2.4. This is why S should contain all the cuts of at least our desired degree of clarity, not just some of them.

To visualize this, suppose our canvas consists of a ring of blue pixels whose interior is filled with random pixels, and which has a white exterior (Figure 6.5). The regions in this picture that our tangles should detect are the ring, and perhaps its white surroundings, but not its interior made up of random pixels. To enable our tangles to identify precisely these two regions we need that S contains the two partitions whose dividing lines are boundaries of our blue ring, inside and outside, and in addition all the partitions that cut out a part of the ring's interior filled with random pixels. If, moreover, S is restrictive enough not to contain partitions that cut through the blue ring or through its white exterior, the tangles of S will correspond to precisely those two regions.

As we shall see in Section 7.4, it is quite common for tangles to be considered simultaneously for different sets S of partitions that conform

FIGURE 6.5.   Two regions, not three, with four yellow lines from S.

to different standards of clarity. Tangles for such sets $S_1 \subseteq S_2 \subseteq \ldots$ of cuts of decreasing standard will have tangles $\tau_1 \subseteq \tau_2 \subseteq \ldots$, which in our case will correspond to regions $R_1 \supseteq R_2 \supseteq \ldots$ of the image with decreasingly clear boundaries.[9] This will be explored further in Chapter 14.

    The point of using tangles of such sets S of cuts, rather than using these cuts directly for image segmentation, is that tangles can enhance the quality of the regions captured in this way. Even from cuts of moderate quality we can get regions of higher quality, since each region is the result of combining many such cuts: those whose orientations, together, form a tangle.

6.3 E-commerce: identifying customer and product types

The context of economics offers a wide range of interesting choices for our set V of 'objects' studied together with some potential 'features' listed in a set S. This is illustrated particularly well by a pair of examples described below, which describes two complementary, or 'dual', aspects of the same scenario.

    In our first example, let V be a set of customers of an online shop, and S the set of items sold at this shop. Let us assume that each customer v has made a single visit to the shop, specifying $s \in S$ as $\vec{s}$ if v included s in his or her purchase, and as $\overleftarrow{s}$ if not. We shall think of the specification $v(S)$ of S as v's 'shopping basket', dividing as it does the set of items into those bought and those not bought.

    An arbitrary specification of S, then, is a hypothetical shopping basket, and a tangle of S is something like a *typical* shopping basket for this set of customers.[10] If we mentally identify a customer v with his or her shopping basket $v(S)$, we may also think of a tangle of S as a (hypothetical) *type of customer* whose 'features' are his or her purchases $v(s) = \vec{s}$ and non-purchases $v(s) = \overleftarrow{s}$.[11]

Alternatively, we may think of S, the set of items in our shop, as our set of *objects* (which would normally be denoted by V, but this letter is taken now), and of the customers in V as potential 'features' of these items: a customer v becomes a feature $\vec{v}$ of precisely those items s that v bought, and a feature $\overleftarrow{v}$ of those items s that v did not buy. Each item s then defines a specification $s(V)$ of V, specifying those v that bought it as $s(v) := \vec{v}$, and those that did not buy it as $s(v) := \overleftarrow{v}$. We may think of $s(V)$ as something like the 'popularity footprint' of the item s with the customers in V.

Note that the information encoded here is no different from that described by our earlier setup, where every customer specified each item s as $\vec{s}$ or $\overleftarrow{s}$. We thus have two ways now of describing the same set of customer preferences.[12]

What are the tangles in our second setup? They are tangles, or typical specifications, of V: hypothetical popularity footprints with the customers in V that are typical for the items in S. If we mentally identify an item s with its popularity footprint $s(V)$, we may think of a tangle of V as a *type of item* whose 'features' are its purchasers $s(v) = \vec{v}$ and non-purchasers $s(v) = \overleftarrow{v}$.

Let us look at some examples of both kinds of tangle, and of how they might be used.

Tangles of S are typical (if hypothetical) shopping baskets, collections of items that are typically bought or avoided together. For example, there might be a tangle, or 'typical shopping basket', that contains all the ecological items in our shop but no environmentally harmful ones. Another might contain mostly inexpensive items and avoid luxury ones.[13]

Note that the 'opposites' of these tangles will not normally be tangles. Indeed, a shopping basket full of luxury goods and avoiding cheap items is unlikely to be typical, because customers that like, and can afford, luxury goods will not necessarily shun inexpensive items. Similarly, inverting the ecological tangle will not produce another tangle, since there is no unifying motivation amongst shoppers to buy environmentally damaging goods or to avoid ecological products.

If we personify these tangles as indicated earlier, and think of them as (hypothetical) customers putting together these hypothetical shopping baskets, we could think of our second tangle as the budget-oriented customer type, one that prefers inexpensive brands and cannot afford or dislikes unnecessarily expensive ones, and of the first as the ecological type that prefers organic foods and degradable detergents but avoids items wrapped in plastic. And, crucially, our tangle analysis might throw up some interesting and economically valuable, but unexpected, customer types as well – perhaps one that prefers items showing the picture of a person on the packaging.

Applications of tangles of S might include strategies for grouping goods in a physical shop, or running advertising campaigns targeted at different types of customer.

Tangles of V, on the other hand, are typical popularity footprints, ways of dividing V into fans and non-fans that occur for many items simultaneously. If what interests us about the items in S is mainly how they appeal to customers, we may think of tangles of V as types of items.

For example, assume that the ecological goods in our shop are substantially more expensive than non-green competing goods. Then V splits neatly into 'eco-purchasers', those customers likely to buy green items whatever their price, and all other customers, who will avoid them because they are more expensive. This division of V, or hypothetical popularity footprint, is borne out in sufficient numbers by the green items in S to form a tangle of V of high agreement, since these items are both liked by the eco-customers and disliked by the others.

An application of finding tangles of V could be to set up a discussion forum for each tangle and invite the fans for that tangle to join. Since they have shown similar shopping tastes, chances are they might benefit more from hearing each others views than could be expected for an arbitrary discussion group of customers.

One last remark, something the reader may be puzzling over at this point. Since our two types of tangle are 'dual' to each other, it so happens that guiding sets for the first type are the same kinds of sets as tangles of the second type: they can both be thought of as sets of customers. But they are not the same. A set $X \subseteq V$ guiding a tangle of S is a set of customers such that for every item $s \in S$ either most of the people in X bought s (if the tangle specifies s as $\vec{s}$) or most of the people in X did not buy s (if the tangle specifies s as $\overleftarrow{s}$). By contrast, a tangle of V (thought of as the set of those v which it specifies as $\vec{v}$)[14] is a group of customers such that every three of them jointly bought some sizeable set of items (at least n).[15]

Put more succinctly: in the first case, most people in some set of customers (not too few) agree about every single item, while in the second case the similarity of the tastes of every few (e.g., three) customers is exhibited by some common set of items (not too few).

6.4 Tangles of words: identifying topics, genres, authors

Training computers to compare texts is a well-established topic in computer science. A number of methods have been suggested that use various criteria to gauge the similarity of two given texts, ranging from word counts to syntax comparisons. When these 'metrics' are used for standard clustering algorithms, they will find corresponding clusters in a given set V of texts: sets of texts that are more similar to each other than to the other texts in V.

The same criteria can be used to define 'features' of the texts in V, among which we can then look for tangles. For example, the particularly frequent use of a certain word, or of a grammatical construction, could be such a feature. A tangle, then, will be a collection of features typical for some of the texts in V. In other words, tangles will be *types* of text, not sets of (similar) texts themselves.

The set S of potential features to look for will depend on our collection of texts: features of syntax that help us compare novels, or novelists, may be less useful when we analyse computer manuals, and word counts for flowers may help with identifying gardening books amongst texts about hobbies but not when we try to classify legal texts. If we already know something about the texts in V, or if we are interested in finding types of a particular kind, we may wish to design S accordingly.

But meaningful collections S of potential features can also be computed mechanically. For example, we might identify some words that occur considerably more frequently than average, either in a particular text or in the texts in V as a whole. For such words we could create a potential feature s, and denote as $\vec{s}$ (say) the feature that this word does occur particularly often, and as $\overleftarrow{s}$ the feature that it does not.

With S chosen, or computed, in this way, we would expect its tangles to capture types of text that can be defined in terms of these features. Thus, if many of our features correspond to the frequent use of particular words, we would expect tangles of S to reflect the topic of a text. If they are chosen to correspond to particular syntactic patterns, they might reflect the text's genre. And if our pre-processing is used to identify phrases used particularly often, they might correspond to authors.

To be sure, such tangles of words[16] will never be what we called 'popularity-based' in Section 1.1: they will not be guided by subsets of V, and thus never be 'clusters' in the sense of Section 6.1. For example, even for a collection of texts all about dogs it is unlikely to happen that each of the typical dog words such as 'heel', 'retrieve' or 'leash', appears in more then half of those texts.

Things look a little better when we consider 'consensus-based' word tangles. In our context, these would be collections of words such that every three of them occur together in at least one of our texts. However, the following example indicates that this is unlikely to work as intended just out of the box.

Suppose V is a collection of texts about hobbies. Our hope is that tangles can identify these hobbies as types of text – including, perhaps, some surprises that identify combinations of interests as hobbies that no-one had thought of. If there are enough texts about gardening in our collection V, we would expect gardening to appear as a tangle. Formally, this tangle should be a specification of S that includes the use of words such as 'rose', 'lawn' or 'watering' as features $\vec{s}$ and words such as 'yeast', 'flour' or 'pre-heated' as non-features $\overleftarrow{s}$.

However, unless our collection V of texts is truly monumental, it is likely to happen that, at least for some triples of typical gardening words such as 'rose', 'lawn' or 'watering', there is no text in V that contains these particular three words. The reason is simply that the set of typical gardening words is too diverse: there is no 'core' of them such that, for any three of those words, we would find in every reasonably large collection of gardening texts one that contains these three.[17]

Does this mean that we will be unable to find a gardening tangle in our collection of texts on hobbies? Not quite – just not the obvious one that chooses *all* the typical gardening words and none of the others. To see this, let us start not by guessing this tangle and then wondering whether our texts support it, but conversely try to build a tangle from whatever texts we have. If our collection does contain texts about gardening, as we assume, let us start by picking three typical gardening words, s_1, s_2, s_3 say, that happen to occur together in one of those texts. Now let τ be the specification of S that orients exactly these three s_i positively, as $\vec{s_i}$, say, for $i = 1, 2, 3$, and all the others negatively. Will this be a tangle of S?

Chances are that it will be. Indeed, consider any set $S' \subseteq S$ of at most three words in S. In order to show that τ is a tangle, we need to find a text $v \in V$ that contains the words from S' (if any) that happen to be among s_1, s_2, s_3 but does not contain any of the other words in S'. If $S' = \{s_1, s_2, s_3\}$, we can use as v the text from which we picked our three words. If S' contains only two words from $\{s_1, s_2, s_3\}$, or even just one or none, it should also be possible to find such a text v, because it is so much easier to find a text that does *not* contain a particular word than to find one that contains it. If this really fails, we can still try other choices of s_1, s_2, s_3 as three typical gardening words occurring together in some text in V, and check whether the specification of S that chooses exactly those three is a tangle.

Our problem, then, is more likely that we get more than one such tangle, perhaps one for every choice of three typical gardening words. And worse still, we will probably get tangles for many arbitrary choices of three words from one of the texts in V too. In Chapter 14, Section 14.8, we shall discuss some other ways of defining S, still based on single words as features of texts but not seeking to specify just these to compute tangles. That approach will help us construct text tangles to identify topics, genres, and even authors, after all.

6.5 Semantics: teaching computers meaning

Our notions determine how we see the world. They are our way of cutting the continuum of our perceptions and ideas into recognizable chunks which, to some extent, persist in time, space, and across different people. Chunks of ideas or perceptions which we bundle together again and again, in different places, and which are similar to the recurring chunks in other people's minds.

Inasmuch as we understand, remember, and communicate aspects of the world around us as structures of notions, the question of what these notions comprise, i.e., which ideas or perceptions we combine into notions, determines what we can understand, remember, or communicate about the world. Since in everyday life[18] we do not choose our notions consciously, understanding and quantifying them becomes significant for our understanding of the world as soon as we seek to compare how people from different cultures or backgrounds understand the world differently.

But as soon as we try to teach a computer what our notions are, perhaps through some interactive process,[19] we need some quantitative

definition of a 'notion'. More ambitiously, if we seek to enable computers
to 'understand' the world by themselves, the question of what should be
their own fundamental notions, not necessarily copying ours, will be top
of the agenda. This is because there are 'good' notions and 'bad' ones,
judged by how they help us understand the world.[20]

Good notions cut the continuum of ideas and perceptions along
lines running between phenomena which we would like to distinguish,
whereas bad notions cut across such phenomena and are therefore less
helpful for distinguishing them. This is not unlike the clustering problem
we considered in Section 6.1. So can we teach a computer to find good
notions, possibly unlike ours, in a way similar to our use of tangles of
set partitions to describe clusters?

If we take this approach, the points which we seek to cluster will
be the phenomena we observe. Groups of similar phenomena, however
vaguely delineated from other such groups, will be bundled into a notion.
The bottleneck partitions of the universe V of phenomena will be those
that do not cut right through a cluster: partitions that do not separate
many similar phenomena. See Chapter 9 for how to find these.

This analogy is flawed in two ways. The first flaw is that, as we saw
in Section 5.3, our notions do not correspond clearly or easily to sets, or
'clusters', of phenomena which they merely label. The second problem is
that neither do tangles: as we saw in Section 6.1, it can make sense to use
tangles to find structure in a dataset *instead of* clustering it into subsets.
Fortunately, though, these two flaws appear to cancel each other out:
notions and tangles are both different from clustering either phenomena
or data points into sets, because they are so similar to each other!

In Section 5.3 we saw how tangles formalize Wittgenstein's idea of
understanding notions as 'family resemblances'. Our tangles there were
specifications of predicates: those that might be offered by a traditional
dictionary in an attempt to define a given notion. While this proved our
point at the time, two aspects of those tangles remained unsatisfactory.

The first of these was an inherent possibility of circularity: notions
were described by tangles of other notions, or predicates. While this did
not invalidate every possible instance of capturing a notion by a tangle
of other notions, it seemed problematic from a philosophical point of
view to seek to establish such definitions as the rule.

The second unsatisfactory aspect of those tangles was that they
were specifications of sets of predicates each tailored to the one notion
we sought to define. We took that list from a hypothetical dictionary,

and merely argued that a *tangle* of those predicates would define the intended term better than a mere conjunction, which was the dictionary's implicit understanding. If we now try to teach a computer our notions we could, of course, allow it a glimpse at such a dictionary. But this would still give us a set of isolated tangles of different sets of predicates, which would leave us unable to apply any mathematics of tangles to the result: the tangle theorems we shall meet in Chapter 8 will all be about tangles of some common set of partitions, or potential features, of a given set V. Moreover, if we hope to enable our computer to come up with its own notions as it seeks to understand the world, there will be no dictionary to appeal to for a list of predicates whose tangles it might form.

The clustering approach from Section 6.1, however, gives us tangles that have neither of these two problems: the set S of partitions of our set V of phenomena can now contain arbitrary such partitions, not just partitions that are themselves defined by notions or predicates; and we can use the same set S for all the notions we wish to capture as tangles.

In summary, tangles appear to be ideally suited to any machine-learning approach to semantics. They differ from clustering in exactly the way in which semantics differs from clustering, while the two appear to offer a perfect match.

Part III

The Mathematics of Tangles

Concepts, theorems, algorithms

This part of the book serves to offer a minimal but precise formal basis for discussing tangles in a mathematical rather than just intuitive context. We begin in Chapter 7 with the main notions around tangles, starting with a brief historical account of their roots in graph theory. We shall see how tangles form a hierarchy in how they describe the structure of a dataset: specific tangles evolve out of fundamental ones, and die again to spawn new tangles – just as small dense clusters form part of sparse bigger ones, or as the detailed notions of 'armchair' and 'dining chair' refine the more basic notion of 'chair', which at their level of detail recedes and becomes less meaningful.

The two main tangle theorems, indicated informally in Chapter 3, will then be explained in Chapter 8 on the basis, and in terms of, the notions made precise in Chapter 7.

In Chapters 9 and 10 we shall see that in setting up tangles to suit a desired application we have quite a bit of leeway. For example, we can influence the hierarchy of tangles mentioned above to suit our intuition of which potential features are more fundamental – and should therefore be specified by most tangles – and which are more specific, and perhaps relevant only in the context of some of our tangles.

In Chapter 11 we describe tangle algorithms that implement the concepts and theorems presented in the earlier chapters: algorithms to compute our 'feature systems' $\vec{S}$, their tangles, and the hierarchy found between these, as well as the tangle-related structures promised by the two tangle theorems.

7

The formal setup for tangles

The purpose of this chapter is to give a more formal definition of tangles: just formal enough to enable readers to try them out in their own professional context if they wish.[1] The definitions here will be less general than is possible, but general enough to model all the examples of tangles we shall be exploring. More general definitions, for which all the theorems still hold, can be found in [9]. For more references on tangles, including recent papers on their theory and applications, please consult my website.

After a brief historical account of how tangles began their life as a concept in graph theory, our first aim in this chapter is to make precise the following notions introduced informally so far:

- *potential features s* of (elements *v* of) *V*

- *features*, or *specifications* $\vec{s}$, $\overleftarrow{s}$ of such *s*

- *consistency* of features

- *tangles*.

Readers consulting the mathematical papers on tangles referenced here should be aware that, for reasons both historical and mathematical, the above terms have different names there, and the notion of tangles is more general. Also, the notions of $\wedge$ and $\vee$ have their meanings swapped.

7.1 Tangles in graphs: how it all began

This section offers a brief historical account of the origin of the notion of tangles in graph theory. It is written for readers familiar with basic graph-theoretic concepts as defined in [7], and it is not needed elsewhere in this book, except occasionally in examples.

In the connectivity theory of graphs and networks one is sometimes interested in identifying the various highly connected regions in a graph, and in how these can be separated from one another in a way that cuts the graph up into chunks. These chunks can then be related to each other in a way that displays the graph's overall structure. For example, every graph decomposes naturally into its connected components, and every connected graph decomposes uniquely into its maximal 2-connected subgraphs and single edges joining them, which together form an overall tree-like structure [7, Figure 3.1.2].

For higher connectivity the picture is less clear, and there are several reasonable notions of 'highly connected region'. These include large complete minors or topological minors, but also k-blocks for large k, or large grids and grid minors. In their seminal work on graph minors [31], Robertson and Seymour noticed that some significant common essence of all these notions can be captured indirectly, as follows.

Consider in a graph G any 'region' H that deserves being called 'highly connected' for some reason, such as any of the above examples. Now consider a separation $\{A, B\}$ of G that has *low order*: one for which the set $A \cap B$ of vertices which separates A from B in G is small. Whatever H is exactly, one thing is clear: it will lie mostly in A or mostly in B. Indeed, if H was spread about evenly over A and B then, since $A \cap B$ is a small separator, H would not deserve to be called 'highly connected'. The highly connected region at the left end of the graph G shown in Figure 7.1, for example, is not divided about evenly by any separation of G of order less than 3: for every such separation it lies almost entirely on one side.

In this way, every highly connected region in a graph G *orients* every low-order separation of G towards one of its two sides, the side that contains most of it. We may think of these oriented separations, all of them collectively, as the 'footprint' of that highly connected region in G.

Such a footprint of a highly connected region H in a graph G reflects only a small part of the information encoded in the actual structure H. But this portion turned out to be just the part of the information that mattered for the structure theory of graph minors that Robertson and

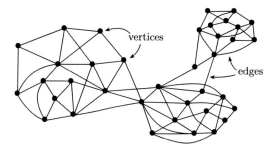

FIGURE 7.1.   A graph G with three highly connected regions.

Seymour were building: the rest, such as how exactly the vertices of H are connected in G – whether by edges or long disjoint paths, for example – is much less relevant and only distracts from its essence. This prompted Robertson and Seymour [32] to give such collective orientations of all the separations of order less than some fixed integer k a name: they called them the *tangles* of G of 'order' k.

The above is not, of course, a usable definition of 'tangle', because the orientations of low-order separations it refers to were in turn defined with reference to some undefined substructure of which we only assumed that it was 'highly connected' in some unspecified sense. This is why tangles in graphs are formally defined in another way: as orientations of all their separations of order less than some fixed integer k such that the union of the 'small' sides of any three such oriented separations, the sides they are *not* oriented towards, is never the entire graph.[2] See [7, Chapter 12.5] for more.

It is easily checked that the footprints of all our earlier examples of concrete highly connected substructures in a graph satisfy this condition, so they are indeed graph tangles. For example, if H is the grid-like region at the left end of the graph G shown in Figure 7.1, then more than three quarters of H – no matter how exactly we define it – lie on one side of any separation of order less than k of G, for any $k \leqslant 4$. Hence if we orient all the separations of G of order less than k towards H, then no three small sides of these oriented separations – the sides on which H does not mostly lie – will together cover all of H, let alone all of G. So H induces a tangle of order k in G.

Thus, all 'highly connected regions' in a graph, no matter how we define these precisely, define a tangle in that graph. Interestingly, it is not known whether the converse of this is true too: whether for every tangle in a graph G, one of order k, say, there exists a set U of vertices of G

such that for every separation of G of order less than k more than half of the vertices in U lie on the side of that separation towards which the tangle orients it – in which case the footprint of U would be exactly this tangle. As we saw in Section 2.4, tangles of set partitions as considered in this book need not have such 'guiding sets' of points.

7.2 Tangles of feature systems

In mathematics, a partition of a set is a collection of disjoint subsets whose union is the whole set. The partitions considered in this book are usually partitions into just two sets, which we assume to be non-empty. Unless it is clear from the context that we are considering more general partitions, a *partition* of a set V in this book is set, or 'unordered pair', $\{A, B\}$ of two disjoint non-empty subsets of V whose union is V. Its two elements, the complementary sets A and B, are its *sides*. The sets A and $B = V \smallsetminus A$, which we also denote by $\overline{A}$, are *inverses* of each other.

When we specify one of the two sides of a partition $s = \{A, B\}$, we say that we *orient* it *towards* that side, and *towards* the elements of V which this side contains. The two sides of s are its two *orientations*, or *specifications*, and we denote them as $\vec{s}$ and $\reflectbox{$\vec{s}$}$.[3]

We thus have three different names for the two sets A and B in a partition $\{A, B\}$ of V: they are the partition's *sides*, its *orientations*, and its *specifications*. We use any one of these, or a mixture, depending only on the context: formally, they mean the same thing.

For example, V might be a set of people whom we asked a yes/no question. Then V is made up, in total, of the set of the people that answered yes and the complementary set of the people that answered no. One of these sets is called A, the other B, but it does not matter which is which. The partition $s = \{A, B\}$ records the information of how the vote was split across V, but is oblivious of which of its two elements corresponds to which vote. This is recorded not in the partition itself but in its two specifications, A and B. So we may think of s as the question, and of $\vec{s}$ and $\reflectbox{$\vec{s}$}$ as the two answers.

Two sets are commonly called *nested* if one contains the other. Similarly, we call two partitions r, s of V *nested* if they have specifications $\vec{r}$ and $\vec{s}$ that are nested as subsets. Note that this is the case if and only if r and s have disjoint specifications: the smaller of the two nested ones is disjoint from the complement of the larger. Partitions of V that are not nested are said to *cross*. For example, the two left-most partitions in Figure 1.3 cross but are nested with all the other partitions indicated.

We write $\vec{S}$ for the set of all the specifications of the elements of a set S of partitions of V. A subset σ of $\vec{S}$ is a *specification*, or *orientation*, *of* S if for every $s \in S$ it contains exactly one of $\vec{s}$ and $\overleftarrow{s}$ (not both). We write $\sigma(s)$ for the one it contains; so $\sigma(s) \in \{\vec{s}, \overleftarrow{s}\}$ for every $s \in S$.

Elements of S are also called *potential features* of (the elements of) V. Elements of $\vec{S}$ are *features*, and $\vec{S}$ itself is a *feature system*.[4] An element of V *has* a feature A if it lies in A. Given two features $\vec{r} = C$ and $\vec{s} = D$, we write

$$\vec{r} \vee \vec{s} := C \cup D \quad \text{and} \quad \vec{r} \wedge \vec{s} := C \cap D.$$

Note that $\vec{r} \vee \vec{s}$ and $\vec{r} \wedge \vec{s}$ need not be elements of $\vec{S}$, in which case they are not called features. But they become features if we add to S the partitions $\{C \cup D, \overline{C \cup D}\}$ and $\{C \cap D, \overline{C \cap D}\}$.

We read $\vee$ as 'or' and $\wedge$ as 'and'. This corresponds to the fact that the elements of V that have the feature $\vec{r} \vee \vec{s}$ (if it is a feature) are those in $C \cup D$, which are precisely those elements of V that have the feature $\vec{r}$ or the feature $\vec{s}$ (or both); and similarly with $\wedge$.

A subset of $\vec{S}$ is *consistent* if every three[5] features $\vec{r}$, $\vec{s}$, $\vec{t}$ in that subset satisfy $\vec{r} \wedge \vec{s} \wedge \vec{t} \neq \emptyset$. Otherwise the subset is *inconsistent*. For example, every feature $\vec{s} = A$ is inconsistent with its inverse $\overleftarrow{s} = \overline{A}$, because $\vec{s} \wedge \overleftarrow{s} = A \cap \overline{A} = \emptyset$. Similarly, if $\vec{S}$ contains features $\vec{r}$, $\vec{s}$ and the inverse of $\vec{r} \wedge \vec{s}$, then these three features form an inconsistent triple.[6] Figure 7.2 shows one consistent and two inconsistent orientations of S.

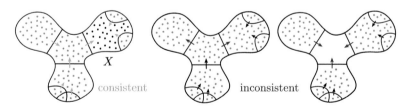

FIGURE 7.2.   One consistent and two inconsistent orientations of S.

Given a set $\mathcal{F}$ of subsets of $\vec{S}$, a consistent specification of S that has no subset in $\mathcal{F}$ is an $\mathcal{F}$-*tangle* of S. A *tangle* of S is an $\mathcal{F}$-tangle for some $\mathcal{F}$ specified in the context. If no $\mathcal{F}$ is specified it is assumed to be empty, in which case tangles of S are just arbitrary consistent specifications of S.

The simplest tangles are the *principal* ones associated with any fixed $v \in V$, the tangles

$$\tau_v = \{ \vec{s} \in \vec{S} \mid v \in \vec{s} \}.$$

These exist for every S and v, so their existence does not tell us anything about S – which, of course, is our aim in studying its tangles.

If S is the set of *all* partitions of V, the principal tangles τ_v are *focussed on* their v: they contain $\vec{s} = \{v\}$, since this lies in $\vec{S}$. They are the only tangles of such S; we proved this in Section 2.4 and will see another proof in Section 7.3. For this reason, we shall not normally be interested in tangles of all the partitions of V.

We often work with $\mathcal{F}$-tangles for $\mathcal{F} = \mathcal{F}_n$, defined as follows. Given an integer $n \geqslant 1$, we write $\mathcal{F}_n$ for the set of all sets $\{ \vec{r}, \vec{s}, \vec{t} \}$ of up to three features that satisfy $\vec{r} \wedge \vec{s} \wedge \vec{t} = U$ for some $U \subseteq V$ with $|U| < n$:

$$\mathcal{F}_n = \{ \{A, B, C\} \subseteq \vec{S} : |A \cap B \cap C| < n \}.$$

Thus, a specification τ of S is an $\mathcal{F}_n$-tangle if for every triple of features in τ there are at least n elements of V that share these three features; note that, since $n \geqslant 1$, such specifications of S are consistent. In fact, the $\mathcal{F}_1$-tangles of any S are precisely its consistent specifications.

In Parts I and II we called these $\mathcal{F}_n$-avoiding specifications of S *typical* for V, the more so the larger n. From now on we shall call them $\mathcal{F}_n$-tangles, or just tangles, in any formal context. But informally we may continue to refer to them as 'typical specifications' of S, or 'types'.

Every $\mathcal{F}_n$-tangle of S contains, among others, all the features $\vec{s}$ for which $\overleftarrow{s}$ has fewer than n elements: this is because it cannot contain $\{\overleftarrow{s}\} \in \mathcal{F}_n$ as a subset, and hence cannot contain $\overleftarrow{s}$ as an element, but must contain either $\overleftarrow{s}$ or $\vec{s}$. Note that n may be a defined in a way that refers to $|V|$ or $|S|$; we then say that n 'is a function of' $|V|$ or $|S|$. As n grows, the requirement for τ to avoid the sets in $\mathcal{F}_n$ gets harder, so we have fewer $\mathcal{F}_n$-tangles for n large.

If the n elements of V that any three features in an $\mathcal{F}_n$-tangle τ must have in common can always be found in some fixed subset X of V, we say that X *witnesses* that τ is an $\mathcal{F}_n$-tangle. Such a set X is indicated in Figure 7.2 for the tangle shown on the left.

The variable n in the definition of $\mathcal{F}_n$ is its *agreement parameter*. The maximum n such that an $\mathcal{F}$-tangle, for any given $\mathcal{F}$, is also an $\mathcal{F}_n$-tangle is its *agreement* value.

7.3 Why triples?

One may reasonably ask why, in the definition of a tangle, we make some consistency requirements of all the subsets of up to three features in that tangle but not, say, of its pairs or quadruples of features.

Suppose, for a moment, that we require consistency only of pairs of features. Such *pre-tangles*, as we might call them, exist for many more feature system $\vec{S}$ than tangles do. In fact, unlike in the case of tangles, we can extend any given set of features that contains no inconsistent pair to such a pre-tangle. To see this, consider any $s \in S$ for which our collection σ of pairwise consistent features contains neither $\vec{s}$ nor $\overleftarrow{s}$. If we cannot add $\vec{s}$ to σ, then this is because it would create an inconsistent pair $\{\vec{r}, \vec{s}\}$ in σ: there must be some $\vec{r} \in \sigma$ such that $\vec{r} \cap \vec{s} = \emptyset$. If we cannot add $\overleftarrow{s}$ to σ, then this is because it would create an inconsistent pair $\{\vec{t}, \overleftarrow{s}\}$ in σ: there must be some $\vec{t} \in \sigma$ such that $\vec{t} \cap \overleftarrow{s} = \emptyset$. But if both these happen then $\vec{r} \subseteq \overleftarrow{s} \subseteq \vec{t}$, which means that $\vec{r}$ and $\vec{t}$ are inconsistent elements of σ, contradicting our assumption.

The fact that pre-tangles exist of so many feature system may not, at first, look like a bad thing. But tangles are intended to describe phenomena that do not always exist: clusters in datasets, or types of features – collections of features often found together. If our data is sufficiently random, or 'noisy', it contains no such clusters or types. So at least some pre-tangles, those that exist even in such noisy data, will not describe what tangles are intended to capture. This does not mean that all pre-tangles are of doubtful value, only some; but those that are valuable are, or at least include, the real tangles.

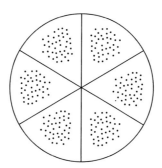

FIGURE 7.3. The pie example for $|S| = 3$: every orientation of S
is a pre-tangle, but only the clusters define tangles of S.

If this sounds a bit philosophical, consider the following quantitative example from Figure 7.3; let us call it the *pie example*. Let S consist of n straight lines through the centre of a disc. These divide the disc into $2n$ pie segments. Place a cluster of points well inside each of them, and let V be the union of these $2n$ clusters. Every $s \in S$ partitions the disc, and hence V, into two sets $\vec{s}$ and $\overleftarrow{s}$ making up complementary half-discs. Since every pair of non-complementary half-discs has a pie segment in common, and hence one of the clusters, it is a consistent pair in $\vec{S}$. Hence each of the 2^n specifications of S is a pre-tangle.[7] But only $2n$ of these have a cluster to correspond to: those whose elements $\vec{s}$ are all oriented towards some fixed pie segment. As we shall see in a moment, these specifications of S are precisely its real tangles, one for each cluster.

Slightly more generally, consider a large dataset D with n clusters, no matter how exactly these are defined. Let S be the set of all 'bottle-neck partitions' of D, partitions $s = \{A, B\}$ such that every cluster lies either mostly in A or mostly in B. Then every such s also defines a partition of the *set V of clusters*: those mostly in A versus those mostly in B. To simplify things, let us from now on view every $s \in S$ as a partition of V rather than of D: the partition into the sets $\vec{s}$ and $\overleftarrow{s}$. Note that S is the set of *all* the partitions of our set V of clusters.

We proved in Section 2.4 that the only tangles of the set of all the partitions of a set V are the principal tangles τ_v focussed on a single v. (In our context here, these correspond to the clusters in D in a one-to-one fashion, as intended.) Let us re-prove this fact in a more inductive way, which uses at every step the fact that tangles ban inconsistent triples rather than just pairs.

Let τ be any tangle of S; our aim is to find some $v \in V$ such that $\tau = \tau_v$. Pick any $s \in S$, and consider its side $\tau(s) =: A_1$. If this has only one element, v say, then $\tau(s) = \{v\} = \tau_v(s)$. But this implies that τ agrees with τ_v on every other $r \in S$, too: if $v \in \vec{r}$, say, then $\overleftarrow{r}$ and $\tau(s) = \{v\}$ form an inconsistent pair, so $\tau(r) = \vec{r} = \tau_v(r)$ as claimed.

Suppose, then, that A_1 has more than one element. Pick a partition of A_1 into non-empty sets B and C. Since $\vec{S}$ contains all the subsets of V, both these and their complements lie in $\vec{S}$. Since $A_1 \cap \overline{B} \cap \overline{C} = \emptyset$, this triple is inconsistent. As $A_1 \in \tau$, this means that τ contains B or C. In fact, it contains exactly one of them: as $B \cap C = \emptyset$, these two sets are inconsistent, and hence not both in τ. Let us assume that τ contains B.

We can now repeat our earlier argument about A_1 with B instead, calling it A_2, say, and continue to find $A_1 \supsetneq A_2 \supsetneq \ldots \in \tau$ until we get to a single-element $A_n = \{v\} \in \tau$, at which point we can deduce that $\tau = \tau_v$ as earlier.

We have shown that, if S consists of *all* the bottleneck partitions of a dataset D with clusters, then these are captured precisely by the tangles of S, while S can have many more pre-tangles: exponentially many in the number of clusters. While this may suffice as justification for banning inconsistent triples rather than just inconsistent pairs in the definition of a tangle, it is important to remember that for more general S there can also be tangles that do not capture a cluster; think of the black hole tangle introduced in Section 2.4, or the text tangles discussed in Section 6.4. But we shall see in Chapter 8, Section 8.2, that there are never more such 'void' tangles than tangles induced by clusters, and unlike pre-tangles they play an important hub-like structural role.

Finally, on the topic of why in the definition of a tangle we forbid inconsistent triples rather than just pairs, there is the technical issue that each of the fundamental tangle theorems from Chapter 8 fails for pre-tangles rather than tangles. The reasons for this are essentially those just discussed, so we shall not dwell on this further.

Let us now turn to the question of why we do not, in the definition of a tangle, forbid quadruples or even larger sets of features whose intersection is empty. In this way we might, for example, rule out the black hole tangle shown in Figure 2.1. The reason such sets of features are not banned in the definition of a tangle is simply that there is no need to: if we did, the structures so defined would just be a more restrictive variant of what are now tangles. But tangle theory works well with tangles as they are, and it always allows us to consider only some specific types of tangles if we wish. In other words, our theorems as they are imply the corresponding assertions for such more restricted tangles. For this reason, there is no need to be more restrictive at the general level when we state or prove these theorems.

It may also be worth pointing out that banning larger sets of features with an empty intersection would, if they are too large, rule out tangles where they are at their best: when they capture *fuzzy* clusters. Such a cluster $C \subseteq V$ is captured by a tangle of S in the current definition if for every $s \in S$ more than two thirds of C lie on one side: then for every three partitions in S the intersection of their 'big' sides, those on which C mostly lies, will be non-empty, so C guides a tangle of S. But it may well happen that every $v \in C$ lies on the 'small' side of *some* $s \in S$.

Then the intersection of *all* the 'big' sides of partitions in S will be empty, although some two thirds of C lie in each of them. If we ruled out such specifications of S in the definition of a tangle by banning large sets of features with an empty intersection, then tangles would no longer capture such fuzzy clusters.

7.4 The evolution of tangles: hierarchies and order

As noted in Section 7.2, we shall not normally be interested in the tangles of the entire set of all the partitions of a set V, because those are just the principal tangles τ_v focussed on their v. This does not mean that we ignore any particular partitions: we just do not consider them all at once.

When S is any set of interesting partitions of V, possibly all of them, we usually study the tangles of subsets $S_1 \subseteq S_2 \subseteq \ldots$ of S that ultimately exhaust S. The tangles of these S_k can interact in interesting ways, and studying this interaction adds greatly to our understanding of how they, all together, describe the features of V.

To see how this works, note that for $i < k$ every tangle τ_k of S_k *induces* a tangle $\tau_i = \tau_k \cap \vec{S_i}$ of S_i. Indeed, since S_i is a subset of S_k, each of its elements s is specified by τ_k as $\vec{s}$ or $\overleftarrow{s}$, and the set of all these specifications of elements of S_i is consistent (because its superset τ_k is). Similarly, if τ_k is an $\mathcal{F}$-tangle for some $\mathcal{F}$ then so is τ_i, since any subset of τ_i in $\mathcal{F}$ would also be a subset of τ_k in $\mathcal{F}$ (which by assumption does not exist).

Conversely, however, a tangle τ of S_i need not *extend to* a tangle of S_k in this way: there may, but does not have to, be a tangle of S_k whose elements in $\vec{S_i}$ are exactly τ. This is because extending τ to a tangle of S_k requires that we specify, in addition, all the $s \in S_k$ that are not already in S_i. There may simply not be a consistent way of doing that, or a way that avoids creating a subset that lies in our chosen $\mathcal{F}$. We shall see an example in a moment.

Similarly, a tangle of S_i can 'spawn' several tangles of S_k that all extend it.[8] We can thus study how tangles are 'born' at some time i (when they are one of several tangles of S_i spawned by the same tangle of S_{i-1}), then 'live on as' (extend to unique) tangles of S_k for a few $k > i$, and eventually 'die' at some time $\ell > k > i$ (when they spawn several tangles of S_ℓ or extend to none).

The sets S_k are usually defined by assigning to every $s \in S$ a number $|s|$, called the *order* of s (and of $\vec{s}$ and of $\overleftarrow{s}$), so that[9]

$$S_k = \{\, s \in S : |s| < k \,\}.$$

Given S, the tangles of S_k are said to be the *tangles of order k in $\vec{S}$*, or the *k-tangles of S*. Such functions $s \mapsto |s|$ are called *order functions*.

Whether or not such an evolution of tangles corresponds to any meaningful refinement of typical features will depend on how we set up our stratification $S_1 \subseteq S_2 \subseteq \ldots$, or choose an order function of S that gives rise to it. We shall look at various ways of doing this in Chapter 9, but we can indicate the main idea here already.

Ideally, the typical features encoded in a tangle of any given order should refine features encoded in the tangles of lower order that it induces. In a questionnaire environment, S_1 would consist of the most basic questions in the survey, and as k increases, the questions become more and more specific.[10]

In our furniture example from Chapter 2, the tangle encoding chairs might extend to a tangle encoding garden chairs, another encoding dining chairs, and a third encoding armchairs. The potential feature of being collapsible or not may serve to distinguish some garden or dining chairs from others. It might then split the corresponding tangles in two, giving rise to a tangle of camping chairs and another of rigid garden chairs, and similarly for the dining chairs. The armchair tangle would be unaffected by this additional potential feature: armchairs are all non-collapsible, and so the feature of not being collapsible will simply integrate into the existing armchair tangle.

In the DNA example of Section 4.2, the elements of S_k for small k would be found at DNA positions at which closely related species have the same bases, so that differences in the specification of S_k for small k would indicate that the corresponding organisms are only distant relatives.

In the clustering example of Figure 1.3 discussed in Sections 2.4 and 6.1, the set S_k for $k = 1, 2, \ldots$ might consist of the partitions with a straight dividing line of length $< k$ in some appropriate unit. When k is just big enough that S_k contains the partitions at the long bottleneck, those across the handle marked in blue, we have two tangles: a tangle τ_1 orienting all these partitions towards their left side, and a tangle τ_2 orienting them all towards their right side.[11] Let us remember this value of k as i, and let k grow. When it gets big enough that the partitions marked red and green are in S_k, the tangle τ_2 dies, spawning three new tangles τ, one for each blob to the right of the handle. Note that each of these induces $\tau_2 = \tau \cap \vec{S}_i$.

The tangle τ_1 lives a little longer. It extends to unique k-tangles for values of k beyond the birth of the three new tangles on the right, and then dies without spawning any new tangles: as soon as k is big enough that the partitions marked by vertical black lines in Figure 7.4 are in S_k, we cannot orient them all consistently, in a way that extends τ_1, except towards a single element $v \in V$ – in which case they cannot form $\mathcal{F}_n$-tangles for $n > 1$.[12]

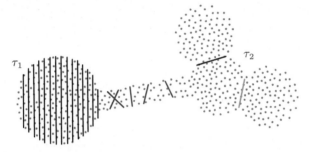

FIGURE 7.4.   For small k, the blue bottleneck gives rise to two
               k-tangles, left and right. As k grows, the right one
               dies when it spawns three new tangles. The left one
               dies a little later, when the leftmost blob is sliced up.

In a more subtle approach than choosing a single stratification $S_1 \subseteq S_2 \subseteq \ldots$ of S we might defer the choice of which potential features to include in S_{k+1} until we have computed the tangles of S_k. In our questionnaire scenario, the question of whether someone likes Tchaikovsky may be relevant when we are investigating how to split a tangle of love for classical music, but less relevant when we analyse a tangle of sport enthusiasm. With this approach, there will be several sets S_{k+1}, one for every choice of a k-tangle (which in turn may be tangles of different sets S_k), and the same potential feature may be given different orders depending on the choice of the tangle relative to which this order is assigned. Our question about Tchaikovsky, for example, might receive order k (so as to be included in S_{k+1}) when we consider the k-tangle of enthusiasm for classical music, but receive some much higher order when we consider a tangle for the love of gardening. On the other hand, when we refine our sports tangle, the Tchaikovsky question may receive some moderately higher order such as $k + 5$: while it will not distinguish any basic sports, it may become relevant to refine a $(k + 5)$-tangle for the love of dance sport into types of which one is classical ballet.

While defining the order of potential features explicitly appears to be the most natural option, there are a few issues we should be aware of. One is that sometimes we simply do not know which potential features are 'basic' and which are not. This may be because our set S is not a hand-designed survey but a collection of technical properties of the objects in V resulting from some measurements whose significance we cannot judge.

A more subtle issue is that by declaring some potential features as more fundamental than others we are in danger of influencing the outcome of our study more than we would like to. Indeed, which tangles of various orders arise in S depends, first and foremost, on how the S_k are chosen. If we are searching for hitherto unknown types of features, we should keep our influence on how to group them to a minimum, unless we know what we are doing and do it deliberately.

Finally, there is a technical problem with choosing the S_k explicitly. This is that our tangle theorems work best for feature systems that are submodular (see Section 7.5): for any $\vec{r}, \vec{s} \in \vec{S_k}$ we would like at least one of $\vec{r} \vee \vec{s}$ and $\vec{r} \wedge \vec{s}$ to be in $\vec{S_k}$ too. We might simply add one of them to $\vec{S_k}$ to achieve this, but this can cause other problems. One is that $\vec{r} \vee \vec{s}$ and $\vec{r} \wedge \vec{s}$, while being well-defined partitions of V, need not lie in S and therefore may lack the 'meaning' that comes with the elements of S, our original potential features. This can make it difficult to decide which of them is the appropriate one to add to S_k because it is more fundamental, or basic, than the other.

In Chapter 9 we shall look at ways to define submodular order functions on all the partitions of a set V generically, rather than explicitly, though still on the basis of a given feature system $\vec{S}$. If some of the potential features in S are more fundamental, or more important, than others, those generically defined order functions can take account of that. In particular, they can be set up so that they assign to partitions s of V that correspond to features, i.e. which happen to lie in S, an order that reflects any pre-assigned importance of s that we may choose: some low order if s is fundamental or important, and high order if s is irrelevant to a particular tangle we are seeking to refine. In this way, those generic order functions can mimic hand-designed ones to some extent, while providing submodularity for free (see Section 9.9).

7.5 Cutting corners: submodularity of features and functions

Consider a feature system $\vec{S}$ and two crossing partitions $r, s \in S$. The four possible intersections of an orientation of r with an orientation of s are the four *corners* of r and s. Figure 7.5 shows two *opposite corners*, $\vec{r} \wedge \vec{s}$ and $\overleftarrow{r} \wedge \overleftarrow{s} = \overline{\vec{r} \vee \vec{s}}$. The other pair of opposite corners of r and s are $\overleftarrow{r} \wedge \vec{s} = \overline{\vec{r} \vee \overleftarrow{s}}$ (top left) and $\vec{r} \wedge \overleftarrow{s} = \overline{\overleftarrow{r} \vee \vec{s}}$ (bottom right).

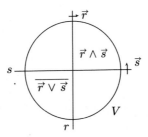

FIGURE 7.5.   Crossing partitions r, s with two marked opposite corners.

The corners of two crossing partitions play a key role in all tangle theorems and advanced algorithms. To understand why this is so, let us take a sneak preview at the first of the two fundamental tangle theorems in Chapter 8, the tree-of-tangles theorem from Section 8.2.

In its simplest form, the theorem claims to find in S a nested subset $T \subseteq S$ that *distinguishes* every pair of tangles of S, in the sense that there exists an $s \in T$ which these two tangles specify differently. Such an s exists in S – otherwise the two tangles would be the same – so the emphasis lies on the claim it can be found in T. This is an ambitious claim, since T is rather a restricted subset of S as it has to be nested.

Without any assumptions about S, such a nested set $T \subseteq S$ clearly need not exist. For example, S might consist of only the two partitions r and s in Figure 7.5. Let us assume that all four corners of r and s are non-empty. Then all four specifications of S are tangles. But the only candidates for our nested set $T \subseteq S$ are the singleton sets $\{r\}$ and $\{s\}$, and neither of these distinguishes all four tangles of S.

The standard way to resolve this problem is to enlarge S. For example, if we add to $\vec{S}$ all four corners of r and s and their inverses, to obtain $\vec{S}'$, say, we still only have four tangles of S', which induce our original four tangles of S. But there is now a nested subset T of S' that distinguishes them all, the set of the four corner partitions.

If a feature system $\vec{S}$ contains all four corners of any crossing $r, s \in S$, it will in fact contain all non-trivial Boolean expressions of its features:

combinations of features built from others by using the symbols $\vee$ and $\wedge$ and taking inverses that are neither empty nor the entire ground set V.[13] However, we do not have to add all corners to $\vec{S}$ to ensure the existence of a nested tangle-distinguishing subset T. In our example it is enough to add any two corners. In general it is enough to ensure that, of any pair of opposite corners of elements of S, one of these corners is also in $\vec{S}$. If S satisfies this, we call it 'submodular'. Theorem 1 in Chapter 8 says that this condition ensures the existence of the desired set T.

Formally, a feature system $\vec{S}$ of partitions of a set V and the corresponding set S of unoriented partitions are called *submodular* if, for every pair $\vec{r}, \vec{s} \in \vec{S}$ such that r and s cross, at least one of $\vec{r} \vee \vec{s}$ and $\vec{r} \wedge \vec{s}$ is also in $\vec{S}$.

Let us check that this definition captures what we intended: that of every pair of opposite corners of $r, s \in S$ one corner is in $\vec{S}$. Let us first consider the two corners indicated in Figure 7.5: the top-right corner $\vec{r} \wedge \vec{s}$ and the bottom-left corner $\overline{\vec{r} \vee \vec{s}}$. To see that one of these lies in $\vec{S}$ if $r, s \in S$ and S is submodular, we apply the submodularity condition to $\vec{r}$ and $\vec{s}$. If it tells us that $\vec{r} \wedge \vec{s} \in \vec{S}$ we are happy, since this is one of our two opposite corners. If not, it will tell us that $\vec{r} \vee \vec{s} \in \vec{S}$. But this is the inverse of the opposite corner. Since $\vec{S}$ is closed under taking inverses, that corner will then be in $\vec{S}$ too, as desired.

How about the top-left and the bottom-right corner? To see that one of these is in $\vec{S}$ we apply submodularity not to $\vec{r}$ and $\vec{s}$ but to either $\vec{r}$ and $\overleftarrow{s}$ or to $\overleftarrow{r}$ and $\vec{s}$. Note that this is covered by the definition of submodularity, since the elements of $\vec{S}$ do not come with fixed names: we are free to call any element of $\vec{S}$ either $\vec{s}$ or $\overleftarrow{s}$ or $\vec{r}$ or $\overleftarrow{r}$ as we please.

If a feature system is not submodular – for example, when S is given as a questionnaire – we can extend S by adding to $\vec{S}$ any missing features and their inverses until it is submodular.

Adding corners to make a given S submodular not only increases S, and thereby our chance to find a nested tangle-distinguishing T in S. Surprisingly, perhaps, it also *decreases* the number of tangles. While this may look counterintuitive at first, since larger sets of partitions have more orientations, it is not hard to see that adding just corners cannot give us more tangles.

The reason is that distinct tangles of a set S' obtained from S by adding to $\vec{S}$ a corner $\vec{r} \wedge \vec{s}$ of $r, s \in S$ and its inverse will induce distinct tangles also on S, because how a tangle orients r and s determines how it must orient this corner: whether it contains $\vec{r} \wedge \vec{s}$ or its inverse. Indeed,

if the tangle contains both $\vec{r}$ and $\vec{s}$ it must contain $\vec{r} \wedge \vec{s}$, since the
triple of $\vec{r}$, $\vec{s}$ and the inverse of $\vec{r} \wedge \vec{s}$ is inconsistent. On the other
hand if the tangle contains either $\overleftarrow{r}$ or $\overleftarrow{s}$, it cannot contain $\vec{r} \wedge \vec{s}$, since
that is inconsistent with both $\overleftarrow{r}$ and $\overleftarrow{s}$. Hence distinct tangles of S' will
never differ only on the corner we added: they must also differ on an
element of S and therefore induce distinct tangles of S. But conversely,
tangles of S need not extend to tangles of S'. This is why S' can, and
often will, have fewer tangles than S.

One may wonder whether the decrease in the number of tangles
resulting from the addition of corners might constitute a loss: might we
be losing tangles that are interesting? Well, the answer is almost always
no: the tangles of S we lose will be tangles only because S is too small to
describe our data adequately, and it will be good to purge our algorithms
of such 'fake tangles'. More on this in Chapter 11.

Historically, the notion of submodularity for feature systems comes
from a notion of submodularity for functions. In our context, we call a
function $A \mapsto |A|$ on the subsets of V *submodular* on $\vec{S}$ if $|A| = |\overline{A}|$ for
all $A \subseteq V$ and

$$|\vec{r} \vee \vec{s}| + |\vec{r} \wedge \vec{s}| \leqslant |\vec{r}| + |\vec{s}| \qquad (7.5)$$

for all $\vec{r}, \vec{s} \in \vec{S}$. Note that $\vec{r} \vee \vec{s}$ and $\vec{r} \wedge \vec{s}$ need not lie in $\vec{S}$ for this
to be the case. An order function on $\vec{S}$ as defined in Section 7.4 will be
called *submodular* if it extends to a function on all the subsets of V that
is submodular on $\vec{S}$.

If S contains all the partitions of V, then the subsystems $\vec{S_k}$ of $\vec{S}$
defined with respect to a submodular order function on S are submodular
as defined earlier: if both $\vec{r}$ and $\vec{s}$ have order less than k then so does at
least one of $\vec{r} \vee \vec{s}$ and $\vec{r} \wedge \vec{s}$, as otherwise they could not satisfy (7.5).

Submodularity is a 'richness condition' we shall have to assume for
the feature systems in our tangle theorems in Chapter 8. It is a necessary
restriction in the sense that without it those theorems would not be true.
But it is not a severe restriction, since many feature systems and order
functions are submodular as they come; see Chapter 9.[14]

Besides, there are interesting applications of tangles that do not
even use our theorems from Chapter 8, and hence do not require submodularity: applications that are content to just discover tangles in the
given data, and to display their hierarchy as discussed in Section 7.4 and
depicted in Figure 14.4.

7.6 Advanced feature systems

Tangles, in this book, are certain consistent specifications of feature systems: to every potential feature they assign one of its two specifications.

The fact that features come in pairs, e.g., as yes/no answers rather than answers with more than two possible values, is an intrinsic property of these tangles. We have argued so far that this does not constitute too much of a restriction, because more flexible answers can be modelled by yes/no answers. For example, suppose a questionnaire Q we wish to evaluate using tangles allows for answers in a range between -5 for 'disagree strongly' to 5 for 'agree strongly'. The simplest way to model this is to think of every question $q \in Q$ as a function $q \colon V \to \{-5, \ldots, 5\}$ that assigns to every $v \in V$ its answer to q. But we can also model the answers given to a question q by virtual yes/no answers to the ten questions of the form 'is $q(v) < i$?', where $i = -4, \ldots, 5$. Features reflecting the results of measurements on a continuous scale of real numbers could be similarly encoded by a finite number of virtual measurements asking whether $q(v) < x$, where x is any of some finite number of thresholds in the range of the possible results of measurement q.

Restricting tangles to binary features like this has one advantage: we can visualize features as subsets of our ground set V – e.g., as the sets of people that answered yes or answered no to a particular question – rather than as partitions of V into many sets which, for different q, may intersect in complicated ways. Similarly we can express the tangle properties, such as consistency, by simple set operations such as requiring the intersection of three sets in the tangle to be non-empty.

This intuitive way of visualising features and their tangles has helped greatly with the development of tangle theory from its origins in graphs, and it remains the best way to think about tangles informally. This is true even when we grapple with tangles mathematically in their most general contexts. Since this book is all about bridging the gulf between the abstract theory of tangles and their down-to-earth potential applications, it is all the more important to keep the formal aspect of tangles restricted to its lowest widely applicable level, which is that of binary feature systems.

However, tangle theory has proceeded far beyond these, and it may be helpful to catch at least a glimpse of the potential range of feature systems and their tangles: although binary feature systems can model all the tangle applications envisaged in this book, others may think of potential applications which, perhaps, cannot be modelled by binary feature systems. To be prepared for this, the tangle software that has been developed so far also works in a more general framework, and readers wishing to adapt this software to their purposes may have an interest in understanding the more general setup.

This section, therefore, introduces *advanced* feature systems and their tangles, systems whose features are real-valued functions rather than binary functions. In order to avoid cluttering our terminology we shall 'update' our existing terms (such as 'feature', 'tangle' and so on) to the more general setup rather than invent a parallel new set of terms. But this will not affect the rest of the book: the new definitions will be used only in the context of advanced feature systems, which are considered mainly in this section and in Chapter 9.

Given a set V, consider the set of all real-valued functions $f\colon V \to \mathbb{R}$. As usual in mathematical notation, we denote this set as $\mathbb{R}^V$. For such functions f, g we write $f \leqslant g$ if $f(v) \leqslant g(v)$ for all $v \in V$. Their *supremum* $f \vee g$ in $\mathbb{R}^V$ is the pointwise maximum of f and g, their *infimum* $f \wedge g$ in $\mathbb{R}^V$ their pointwise minimum.

Note that if f encodes a subset A of V in that it maps the elements of A to 1 and the other elements of V to -1, and g likewise encodes $B \subseteq V$, then $f \vee g$ encodes the set $A \cup B$ while $f \wedge g$ encodes the set $A \cap B$.

Similarly, if we think of two maps $p, q\colon V \to \{-5, \ldots, 5\}$ as recording the answers of the people in V to two questions p and q, then the infimum $p \wedge q$ of these maps is another map that records 'less positive' answers than p and q themselves did. If we intepret the values in $\{-5, \ldots, 5\}$ of these maps as expressing consent, ranging from 'disagree completely' (for -5) to 'agree completely' (for 5) as earlier, this means that $p \wedge q$ expresses lower consent than p and q. This is exactly what we expect of the conjunction of two statements. So reading $p \wedge q$ as 'p and q', the usual way to read the symbol $\wedge$ in mathematics, is borne out by the interpretation of -5 as maximum disagreement and of 5 as maximum agreement to the statement proposed.

The *additive inverse* of a function $f\colon V \to \mathbb{R}$ is the function $-f$ which maps every $v \in V$ to $-f(v)$. For example, an advanced feature that records people's taste for bananas assigns 5 to those that that love bananas and -5 to those that hate them: *not* to the people that simply do not love bananas. The additive inverse of a feature thus expresses something like its opposite, not its negation.

If f encodes a partition of V by sending some elements of V to 1 and the others to -1, then the two notions of inversion – of functions and of partitions – are compatible: just associate a function f with the subset $f^{-1}(1)$ of V.[15]

An *(advanced) feature system* is a set $\vec{S}$ of functions $V \to \mathbb{R}$ that is closed under additive inversion: for every $f \in \vec{S}$ we ask that also $-f \in \vec{S}$. We do not require that $\vec{S}$ must also be closed under taking infima and suprema, although we may wish to add these to $\vec{S}$ to achieve this. The elements of a feature system are *features*. Given a feature $f \in \vec{S}$, the pair $\{f, -f\}$ is a *potential feature*.[16]

Given a feature system $\vec{S}$, we write S for the set of all potential features $\{f, -f\}$ with $f \in \vec{S}$, and call a subset τ of $\vec{S}$ a *specification* of S or $\vec{S}$ if τ contains exactly one of f and $-f$ for every $f \in \vec{S}$. We often denote potential features $\{f, -f\}$ in S by simple letters such as s, and their *orientations*, the features f and $-f$, as $\vec{s}$ and $\overleftarrow{s}$.[17]

Two potential features $r, s \in S$ are *nested* if they have comparable orientations: if there exist orientations $\vec{r}$ of r and $\vec{s}$ of s such that $\vec{r} \leqslant \vec{s}$. If r and s encode partitions of V by mapping it to $\{-1, 1\}$ as earlier, then this corresponds to our earlier notion of nestedness of potential features. Potential features that are not nested are said to *cross*. Collections of pairwise nested potential features are called *tree sets*; we shall explore these for set partitions in Section 8.1.

A feature system $\vec{S}$ is *submodular* if for all $\vec{r}, \vec{s} \in \vec{S}$ at least one of $\vec{r} \wedge \vec{s}$ and $\vec{r} \vee \vec{s}$ is also in $\vec{S}$. When r and s encode crossing partitions of V, as earlier, then this corresponds to our earlier notion of submodularity for feature systems of set partitions. But we now require the above also when r and s are nested.[18]

The notion of submodularity for order functions is the same now as in Section 7.5: a function $|\cdot|$ on $\mathbb{R}^V$ is *submodular on* $\vec{S}$ if $|\vec{s}| = |\overleftarrow{s}|$ and

$$|\vec{r} \vee \vec{s}| + |\vec{r} \wedge \vec{s}| \leqslant |\vec{r}| + |\vec{s}| \tag{7.5}$$

for all $\vec{r}, \vec{s} \in \vec{S}$. As before, $\vec{r} \vee \vec{s}$ and $\vec{r} \wedge \vec{s}$ need not lie in $\vec{S}$ for this to be the case. A function on $\vec{S}$ is *submodular* if it extends to a function on $\mathbb{R}^V$ that is submodular on $\vec{S}$.

A set of features is *consistent* if it contains no triple $\vec{r}, \vec{s}, \vec{t}$ such that $\vec{r} \wedge \vec{s} \wedge \vec{t} \leqslant 0$ pointwise: if for every choice of up to three features in that set there exists $v \in V$ on which all three of them are positive. Once again, this agrees with our earlier notion of consistency of features if $\vec{r}, \vec{s}, \vec{t}$ encode subsets of V by taking values in $\{-1, 1\}$.

As before, the consistent specifications of a feature system are its *tangles*. Note that if τ is a tangle of $\vec{S}$ and $\vec{r}, \vec{s} \in \tau$, then τ must also contain $\vec{r} \wedge \vec{s}$ (as long as this lies in $\vec{S}$), since its inverse $-(\vec{r} \wedge \vec{s})$ forms an inconsistent triple with $\vec{r}$ and $\vec{s}$: for every $v \in V$, if both $\vec{r}$ and $\vec{s}$ take a positive value at v then so does their pointwise minimum $\vec{r} \wedge \vec{s}$, which means that $-(\vec{r} \wedge \vec{s})$ takes a negative value at v. So at every $v \in V$ at least one of $\vec{r}$, $\vec{s}$ and $-(\vec{r} \wedge \vec{s})$ is negative or zero.[19]

If $p, q : V \to \{-5, \ldots, 5\}$ record the answers of the people in V to two questions, as earlier, and p and q are consistent as $V \to \mathbb{R}$ functions, then p and q are consistent also with $p \wedge q$ but not with its inverse $-(p \wedge q)$. This agrees once more with reading $\wedge$ as 'and': if we agree with both p and q we should also agree with $p \wedge q$ (if asked), but should disagree with the opposite of the statement $p \wedge q$.

As usual, tangles that have no subset in some collection $\mathcal{F}$ of subsets of $\mathbb{R}^V$ are called $\mathcal{F}$-*tangles*. For example, we might consider as $\mathcal{F}$ the set $\mathcal{F}_n$ of all sets $\{\vec{r}, \vec{s}, \vec{t}\}$ of features $V \to \mathbb{R}$ such that fewer than n elements of V satisfy $(\vec{r} \wedge \vec{s} \wedge \vec{t})(v) > 0$. The $\mathcal{F}_n$-tangles of S, then, are those of its specifications τ such that every three features in τ take positive values on some n elements of V.

As a variant that takes better note of the values of features $V \to \mathbb{R}$, rather than just of the sign of these values as in the definition of $\mathcal{F}_n$, one might consider as $\mathcal{F}$ the set $\mathcal{F}'_n$ of all sets $\{\vec{r}, \vec{s}, \vec{t}\}$ such that

$$\sum \left\{ \vec{r}(v) + \vec{s}(v) + \vec{t}(v) \mid (\vec{r} \wedge \vec{s} \wedge \vec{t})(v) > 0 \right\} < n.$$

In other words, an $\mathcal{F}'_n$-tangle requires of every triple of features it contains not simply that there are many points v (at least n) on which all three of these features are positive, but that the sum of the values of those v must be large, at least n. For advanced features that encode partitions by taking values -1 and 1 only, the $\mathcal{F}'_{3n}$-tangles are precisely their $\mathcal{F}_n$-tangles. Thus, $\mathcal{F}'_n$-tangles of advanced features generalize our familiar tangles of set partitions as well as $\mathcal{F}_n$-tangles of advanced features do.

Advanced features $f : V \to \mathbb{R}$ can be used to define partitions of V in various ways[20] – for example, into the set of $v \in V$ on which f is positive and the rest of V. The tangles and $\mathcal{F}_n$-tangles defined above for advanced feature systems then induce tangles and $\mathcal{F}_n$-tangles of these partitions. This does not mean, however, that advanced feature systems are redundant. The $\mathcal{F}'_n$-tangles defined above work directly with their values, rather than just the sign of those values. The most important order functions for partitions, see Chapter 9, require that the partitions are given as advanced features. And our choice of an order function forms the basis for the hierarchy of tangles explained in Section 7.4.

7.7 Duality of feature systems

In Section 6.3 we saw an intriguing example of a pair V, S of sets each of which could be regarded as a set of potential features of the other. Each of them had its own tangles, and these tangles described different aspects of the relationship between V and S. Now that we have defined tangles and their ingredients more formally, we can show that such a 'duality' between V and S exists always, not only when their interpretations in the application at hand suggest it.

In other words, whenever we encounter a natural setting for studying tangles, there is always also a dual setting that may have tangles. These may shed additional light on the situation we are studying, and may thus be worth exploring even when this was not initially envisaged. For example, we shall see in Chapter 9 how tangles of the dual system can be used to define order functions on a given feature system. And in Chapter 12, Section 12.4, this duality appears naturally, though quite unexpectedly, in the context of developing drugs targeting a given set of features of pathogens.

Consider two disjoint sets, X and Y. Assume that some elements x of X are in a 'special relationship' with some elements y of Y. We can express this formally by way of a set E whose elements are sets $\{x, y\}$ with $x \in X$ and $y \in Y$: not all of them, but some.[21]

Our choice of E defines for every $x \in X$ a subset $\vec{x}$ of Y consisting of just those y with which x has this special relationship. Similarly, for every $y \in Y$ we have the set $\vec{y}$ of all $x \in X$ with which y has this special relationship:

$$\vec{x} := \{\, y \in Y \mid \{x, y\} \in E \,\} \quad \text{and} \quad \vec{y} := \{\, x \in X \mid \{x, y\} \in E \,\}.$$

Then for all $x \in X$ and $y \in Y$ we have

$$x \in \vec{y} \;\Leftrightarrow\; y \in \vec{x} \qquad\qquad (*)$$

as both these are true if $\{x, y\} \in E$ and false otherwise.

Let us define $\overleftarrow{x}$ as the complement of the set $\vec{x}$ in Y, and $\overleftarrow{y}$ as the complement of the set $\vec{y}$ in X. Further, let

$$\vec{X} := \{\, \vec{x} \mid x \in X \,\} \cup \{\, \overleftarrow{x} \mid x \in X \,\}$$
$$\vec{Y} := \{\, \vec{y} \mid y \in Y \,\} \cup \{\, \overleftarrow{y} \mid y \in Y \,\}.$$

Then $\vec{X}$ is a set of features of Y, and $\vec{Y}$ is a set of features of X, with the partitions $\{\vec{x}, \bar{x}\}$ of Y and $\{\vec{y}, \bar{y}\}$ of X as potential features. As $\vec{X}$ and $\vec{Y}$ are closed under complementation, they are in fact feature systems. We say that these feature systems are *dual* to each other.[22]

Note that every x determines the set $\{\vec{x}, \bar{x}\}$, and similarly every y determines the set $\{\vec{y}, \bar{y}\}$. If these correspondences are 1–1, i.e., if there are no $x \neq x'$ with $\{\vec{x}, \bar{x}\} = \{\vec{x'}, \bar{x'}\}$ and similarly for the ys, we may choose to ignore the formal difference between x and $\{\vec{x}, \bar{x}\}$, and between y and $\{\vec{y}, \bar{y}\}$. Then X becomes the set of potential features of (elements of) Y that underlies the feature system $\vec{X}$, and Y becomes the set of potential features of (elements of) X that underlies the feature system $\vec{Y}$, as usual in our terminology.

Let us now see how in general, given any set S of partitions of a set V, we can define a system $\vec{V}$ of features of S that is dual in this sense to the system $\vec{S}$ of features of V, and which in particular satisfies the analogue of $(*)$. We start by picking for every $s \in S$ a default specification, which we denote as $\vec{s}$ (rather than $\bar{s}$). This determines for every $v \in V$ the sets

$$\vec{v} = \{ s \in S \mid v \in \vec{s} \} \quad \text{and} \quad \bar{v} = \{ s \in S \mid v \in \bar{s} \},$$

which form a partition of S. If, as we shall assume, these partitions $\{\vec{v}, \bar{v}\}$ differ for distinct $v \in V$, they determine their v uniquely and we may think of each v as shorthand for $\{\vec{v}, \bar{v}\}$. This makes V into a set of partitions $v = \{\vec{v}, \bar{v}\}$ of S and

$$\vec{V} := \{ \vec{v} \mid v \in V \} \cup \{ \bar{v} \mid v \in V \}$$

into the corresponding feature system. Then $\vec{V}$ and $\vec{S}$ form an instance of a pair $\vec{X}, \vec{Y}$ of dual feature systems as defined earlier.[23]

With this setup in place, one can investigate how the tangles of S are related to those of V [11]. Note that the tangles of V will depend not only on V and S, but also on the default orientation we chose for S.

For advanced feature systems there is a similar duality, which is even easier to state. Indeed, given a system $\vec{S}$ of advanced features $\vec{s}: V \to \mathbb{R}$ as discussed in Section 7.6, we define $S \to \mathbb{R}$ maps $\vec{v}$ and $\bar{v}$ for all $v \in V$ by letting

$$\vec{v}(s) := \vec{s}(v) \quad \text{and} \quad \bar{v}(s) := \bar{s}(v) \qquad (**)$$

for all $s \in S$. Then $\bar{v} = -\vec{v}$ for all v, so

$$\vec{V} := \{ \vec{v} \mid v \in V \} \cup \{ \bar{v} \mid v \in V \}$$

is an advanced feature system that is dual to $\vec{S}$ in the sense of $(**)$. As before, if the sets $\{\vec{v}, \bar{v}\}$ differ for distinct $v \in V$, they determine their v uniquely and we may think of each $v \in V$ as shorthand for $\{\vec{v}, \bar{v}\}$. Then V becomes the set of potential features of the feature system $\vec{V}$, as usual.

We have now seen enough of this duality to apply it in practice: whenever we have a set V of 'objects' at hand, and a set $\vec{S}$ of features of these objects, then together with investigating the tangles of S we can also investigate the tangles of V viewed as a set of potential features of the elements of S. Together they will paint a more comprehensive picture of the situation we are trying to analyse than either of the two types of tangle alone can offer.

8
Tangle theorems

In this chapter we present the two most fundamental theorems about tangles, the *tree-of-tangles theorem* and the *tangle–tree duality theorem*, in their simplest forms. The versions presented here are a far cry from their most general versions, but strong enough to apply to all the application scenarios we have discussed as examples, and likely to apply to most application scenarios that arise in the empirical sciences.

The tree-of-tangles theorem, of which we present two versions in Section 8.2, finds in a given set S of potential features a small subset T that suffices to distinguish all the tangles of S: for every two tangles of S there is a potential feature in T, not only in S, which these two tangles specify differently.

In fact, T is small for a structural reason: its elements are pairwise nested. Such sets are called *tree sets*. Tree sets of partitions of a set structure this set in such a way that it resembles the shape of a tree. When tangles of S are induced by clusters in V, our tree set $T \subseteq S$ structures V in such a way that these clusters lie, essentially,[1] in the local, knobbly, bits of this tree shape when we view it as in Figure 8.1. But tangles of S also reside in these 'locations' of such a tree set $T \subseteq S$ if they are not induced by a cluster, such as the black hole tangles from Section 2.4. We look at tree sets and their own tangles first, in Section 8.1, before we get to the tree-of-tangles theorem in Section 8.2.

In Section 8.3 we present the tangle–tree duality theorem. If S has no tangle of a desired type, this theorem finds a small subset T of S that suffices to demonstrate this fact conclusively. We say that T demonstrably 'precludes' the existence of a tangle. This can help us prove that some given data is, in a precise and quantitative sense, unstructured or contaminated.

In Section 12.1 we will discuss why the tangle-distinguishing property of the set T found by the tree-of-tangles theorem makes it particularly valuable: we can predict from how a given $v \in V$ specifies the potential features in T how it will probably specify most of the other $s \in S$.

8.1 Tree sets of partitions

A collection of pairwise nested partitions of a set V is a *tree set*. While arbitrary sets of partitions of V can be exponentially large in terms of $|V|$, tree sets of partitions are much smaller, smaller than twice $|V|$. The tangle-distinguishing tree sets T that we find in Section 8.2 will be smaller still, no larger than the number of tangles they distinguish.

The fact that tree sets of partitions are so small is a consequence of their deeper structural properties, which are even more important for us than size alone. These structural properties also imply that tree sets can be nicely drawn in the plane as in Figure 8.1, so as to resemble a tree.

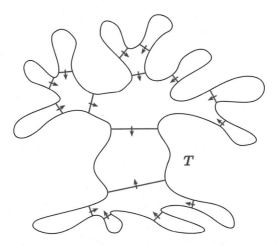

FIGURE 8.1.   A tree set T of 'bottleneck partitions' of a set V, with arrows indicating a tangle of T at a central location.

Why is this so rare and surprising? When we draw a single partition $\{A, B\}$ of a set V on a piece of paper, we choose where we draw the points from V according to this partition. For example, we start by drawing a disc to put the points in, place all the points in A to the left of all those in B, and draw a vertical line between A and B to indicate this partition. We can still do this for two crossing partitions, as in Figure 7.5: there, both partition are marked by a line between boundary points of the disc, one dividing its sides as left/right, the other as top/bottom. For larger sets of arbitrary partitions of V, however, this may no longer be possible: their sides will intersect in ways too complicated to accommodate them all so cleanly in the same picture..

Tree sets of partitions of V, however, can always be displayed in this way: by placing the points of V in a disc so that each of the partitions divides that disc into two connected parts by a line segment between two boundary points, and even so that these dividing lines do not cross.[2]

This is a non-trivial mathematical statement which we cannot prove rigorously here.[3] But it is important to remember: it underlies much of our intuition for tree sets, which in turn will be needed to visualize all the tangle theorems we shall meet in this chapter.

The tangles of a tree set are also much easier to visualize than those of an arbitrary feature system. Indeed, if we place an arrow on each of the partitions forming the tree set T in Figure 8.1 to indicate how some fixed tangle τ orients them, then no two of these arrows will point away from each other: if they did, then these two oriented partitions would form an inconsistent pair in τ, contradicting our assumption that τ is a tangle. Following these arrows in the picture we will therefore arrive at a region of the 'tree' to which *all* the arrows drawn by τ point. We think of this region as the 'location' of T at which τ 'lives'.

Formally, we define the *location* of a pre-tangle[4] τ of a tree set as the set of its minimal elements: those which, when viewed as subsets of V, are minimal in τ with respect to inclusion.[5] We say that τ *lives*, or is at *home*, at its location. The *locations* of a tree set are the locations of its pre-tangles.

We can recover any pre-tangle τ of a tree set T from its location σ: given any $t \in T$, only its specification in τ is consistent with σ.[6] Thus, once we know the location of τ, the set of its minimal elements, we know the entire τ. Different pre-tangles of T thus live at different locations.

If T consists of partitions of V, and τ is a pre-tangle of T with location σ, then clearly

$$V_\tau := \bigcap \{\, A \mid A \in \tau \,\} = \bigcap \{\, A \mid A \in \sigma \,\}.$$

We then also say that the elements of V_τ *live at* this location σ of T.

Every $v \in V$ lives at some location of T: at that of its tangle τ_v of T. And no $v \in V$ can live at more than one location, since $V_\tau \cap V_{\tau'} = \emptyset$ for distinct pre-tangles τ, τ' of T.[7] The non-empty sets V_τ, where τ ranges over all the pre-tangles of T, therefore partition V in the usual more general sense of the word.

There can, however, be locations of T at which no v lives: locations of pre-tangles τ of T with $V_\tau = \emptyset$. We call such locations *empty*. The black hole tangle from Section 2.4 is simple example.

Locations of pre-tangles of tree sets are examples of stars.[8] A *star* is a set of features whose inverses are disjoint subsets of V.[9] Equivalently, a set of features forms a star if each of them contains the inverses of all the others. It is easy to check that these assertions hold for locations of pre-tangles of tree sets,[10] so these are examples of stars. The sets $\mathcal{F}$ that we use to define tangles will also often consist of stars.

Stars also arise naturally as in the following example. Consider a survey S whose questions allow answers on a scale from -5 to 5. To compute its tangles, we may wish to convert S into a hypothetical questionnaire of yes/no questions that is equivalent to S in the sense that its answers can be computed from the (known) answers recorded for S, and whose (computed) yes/no answers conversely imply the original answers to S. One way to do this is to replace each $s \in S$ with eleven questions $s_{-5}, \ldots, s_5$ asking whether the answer given to s is $-5, \ldots, 5$, respectively, yes or no. For $i = -5, \ldots, 5$ let $s_i = \{A_i, \overline{A}_i\}$, where $\vec{s}_i = A_i$ consists of the yes-answers. The sets A_i are disjoint, because every $v \in V$ had to choose only one of the eleven possible answers to s, which makes it an element of only one A_i. So the $\overline{A}_i$ have disjoints complements, which means that all the $\overleftarrow{s}_i$ together form a star.

8.2 Tangle-distinguishing feature sets

Consider two different tangles of a set S of partitions of a set V. As both our tangles are specifications of S, the fact that they are different means that there exists an $s \in S$ which one of them specifies as $\vec{s}$, and the other as $\overleftarrow{s}$. We say that s *distinguishes* the two tangles. More generally, we say that a subset T of S *distinguishes all the tangles of S* if for every two tangles of S we can find in this set T an s that distinguishes them.

Our first theorem says that, if S is submodular, we can find such a tangle-distinguishing subset T of S that is a tree set:

Theorem 1. (Tree-of-tangles theorem for fixed S [12])
*For every submodular feature system $\vec{S}$ there is a tree set $T \subseteq S$ that
distinguishes all the tangles of S.*

Figure 8.2 shows two tangles of S distinguished by a partition $t \in T$.
Informally, we call the tree set T in Theorem 1 a *tree of tangles*, or more
specifically, a *tree of the tangles of S*.

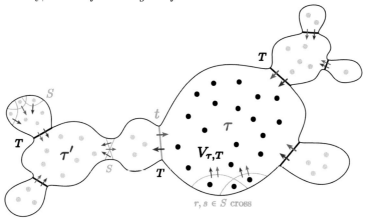

FIGURE 8.2.   Tangles τ, τ' of S living at different locations of T.

Theorem 1 also holds for the advanced feature systems discussed
in Section 7.6. Both versions have a strengthening that provides a
'canonical' tree set T [17]. This means, very roughly, that T reflects
the symmetries inherent in the structure that S imposes on V. More
about this in a moment.

Every tangle τ of S defines a tangle on every subset T of S: the elements of τ that happen to lie in $\vec{T}$ form a tangle of T, because they
contain no inconsistent triple and no set from whatever $\mathcal{F}$ we prescribe
for tangles of S. Let us say that, given a tree set $T \subseteq S$ as in Theorem 1,
a tangle τ of S *lives at* the location of T specified by τ in this way, the
location of $\tau \cap \vec{T}$. We write

$$V_{\tau,T} := \bigcap \{\, A \mid A \in \tau \cap \vec{T} \,\}$$

for the set of all $v \in V$ living at this location (Figure 8.2).

Different tangles of S live at different locations of the tree set T
provided by Theorem 1, because T distinguishes them.[11] Some of these
locations can be empty, in which case they may or may not be home to

a tangle of S.[12] If T is chosen minimal, then all its locations are home to a tangle of S, even if they are empty.[13]

To illustrate this, recall the *black hole tangle* from Section 2.4. There, V contains $n \geqslant 4$ well-separated clusters, and S consists of the n nested partitions of V that each separate one cluster from the others. Consider $n = 4$. Then S has five tangles: one induced by each of the four clusters, and a central 'black hole' tangle that orients each of the four partitions away from 'its' cluster, into the 'void middle'. The tree set T found by Theorem 1 in this example will be all of S: no proper subset of S distinguishes all five tangles, and S is itself a tree set.

Let us now enlarge S by adding a further partition s of V that has two of the four clusters on its left, and two on its right. The four tangles of the original example that were induced by the four clusters clearly extend to tangles of the enlarged system S: just orient s left or right according to where the cluster lies. But the 'black hole' tangle τ that oriented each of the four partitions away from 'its' cluster does not extend to a tangle of the enlarged S: no matter how we orient s, it will form an inconsistent triple with two elements of τ, the two partitions that split off the two clusters towards which we oriented s.

When we apply Theorem 1 to this enlarged S, it could return the tree set $T = S \smallsetminus \{s\}$ that still has five locations, of which the 'central' one is still empty but no longer home to a tangle of S. Since the enlarged S is still a tree set, the theorem could also return $T = S$, although the new partition s is not needed to distinguish any clusters. If it does, then T will have two adjacent empty locations, neither of which is home to a tangle of S.

Now consider the analogous example with only three clusters and three partitions in S. Now there is no central tangle, since orienting the three elements of S away from the cluster they split off yields an inconsistent triple. So now the tree set T from Theorem 1 has to distinguish only three tangles, those corresponding to the three clusters in V.

Choosing $T = S$ is still a possible solution, and in fact it is the unique *canonical* tree set that distinguishes all three tangles (see below). This T has three locations each containing a cluster and home to a tangle of S, as well as a central empty location. This 'black hole', however, is not home to any tangle, not even a black hole tangle.[14]

But note that this $T = S$ is not minimal for the purpose of witnessing Theorem 1: if we omit any one of its three elements, the other two will still distinguish all three tangles of S. However, as the choice of which element of S we delete to form T in this way is necessarily arbitrary, because the three elements of S are entirely symmetrical in how they partition V, this T will never be canonical.[15]

An important theoretical consequence of Theorem 1 is that S has only few tangles, even of the most general kind (with $\mathcal{F} = \emptyset$). Indeed, as T distinguishes all the tangles of S, these induce distinct tangles of T. Now a tree set of size ℓ has at most $\ell + 1$ tangles.[16] And as we noted in Section 8.1, all tree sets in S are small: they have size some $\ell < 2|V|$. So the number of all tangles of S, not just of T, is at most $|T| + 1 \leqslant 2|V|$.

In practice, the implication also goes the other way. By specifying $\mathcal{F}$, we can control roughly how many tangles we get. For example, the larger we choose the agreement parameter n, the larger the resulting family $\mathcal{F}_n$. As $\mathcal{F}$ grows, the number of $\mathcal{F}$-tangles of S decreases, because not containing a triple from $\mathcal{F}$ is harder when $\mathcal{F}$ is larger. Since T, if chosen minimal, is smaller than the number of tangles of S, it will be small whenever we choose $\mathcal{F}$ so that S has few tangles.

This aspect is particularly valuable if S is not submodular (as Theorem 1 requires). To distinguish ℓ tangles we only need $\ell - 1$ elements of S, and these are quick to find. However, they may not form a tree set. Our algorithm[17] will then build from these elements of S a tree set T, also of size $\ell - 1$, which once more distinguishes all our ℓ tangles of S. This T will not be a subset of S. But the elements of $\vec{T}$ will be Boolean expressions of specifications of those $\ell - 1$ elements of S we started with, obtained by iteratively adding features such as $\vec{r} \vee \vec{s}$ or $\vec{r} \wedge \vec{s}$ combined from existing or previously added features $\vec{r}$ and $\vec{s}$. So this T will be small whichever way we choose to measure it: in absolute terms, as $|T|$, or in terms of how many features from $\vec{S}$ are needed to build it.

There is no general rule about how many tangles to aim for when we set the agreement parameter n: this will depend on the application. Choosing n large will give us only the most pronounced tangles, while choosing it smaller will give us more, but less pronounced, tangles.

If S satisfies a little more than submodularity, we can dramatically improve Theorem 1 to find a tree set in S that even distinguishes tangles of different orders, all at once. This will be our next theorem. Assume for this that S is the entire set of all partitions of V, equipped with a submodular order function $s \mapsto |s|$. Recall that any k-tangle of S for some k – i.e., a tangle of $S_k = \{\, s \in S : |s| < k \,\}$ – is a tangle *in* $\vec{S}$. Two tangles in $\vec{S}$, not necessarily of the same order, are *distinguishable* if there exists an $s \in S$ which they specify differently.[18] Such an s

distinguishes the two tangles, and it does so *efficiently* if no $r \in S$ of lower order distinguishes them. A set $T \subseteq S$ *distinguishes all the tangles in $\vec{S}$ efficiently* if for every two distinguishable tangles in $\vec{S}$ there is an $s \in T$ that distinguishes them efficiently.

Theorem 2. (Tree-of-tangles theorem for tangles of variable order [13]) *For every submodular order function on the set S of all partitions[19] of V there is a tree set $T \subseteq S$ that distinguishes all the tangles in $\vec{S}$ efficiently.*

Figure 8.3 shows two indistinguishable tangles in $\vec{S}$: a k-tangle τ that is a subset of a $(k+1)$-tangle τ' refining it. See [13, 20] for more general and canonical versions of Theorem 2, also for advanced feature systems. Informally, we call the tree set T in Theorem 2 a *tree of tangles*, or more specifically, a *tree of the tangles in $\vec{S}$*.

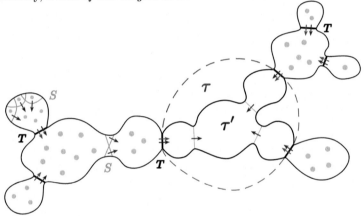

FIGURE 8.3.   A $(k+1)$-tangle τ' of S refining a k-tangle τ.

As earlier, every tangle τ in $\vec{S}$ induces a tangle in $\vec{T}$, the tangle $\tau \cap \vec{T}$. But note that this is no longer a tangle of the entire set T, only of a subset: the set of those partitions in T whose order is less than the order of τ. If the set of minimal elements of $\tau \cap \vec{T}$ is a location of T, we say that τ *lives at* this location of T; otherwise it has no 'home' in T. The k-tangle in Figure 8.3, for example, has no home in the tree set shown, but the $(k+1)$-tangle inducing it does.

As before, the the sets

$$V_{\tau,T} := \bigcap \{ A \mid A \in \tau \cap \vec{T} \}$$

have union V, since every $v \in V$ induces tangles τ_v of all orders. The sets $V_{\tau,T}$ are disjoint for different τ if neither of these τ refines the other, because T distinguishes such tangles. As earlier in the case of Theorem 1, the sets $V_{\tau,T}$ can also be empty. If they are, they may or may not be home to a tangle in $\vec{S}$. The non-empty sets $V_{\tau,T}$ partition V as τ varies over the maximal[20] tangles in $\vec{S}$, those not refined by any other tangle.

The tree shape of a tree set T, as displayed once more in Figures 8.2 and 8.3, can be formalized by a graph-theoretical tree $\mathcal{T}$ whose edges correspond to the elements of T, and whose maximal graph-theoretical stars, the sets of edges at a given node, correspond to the locations of T. In the case of Theorem 2, this tree $\mathcal{T}$ also displays the non-maximal tangles in $\vec{S}$ and which of these refines which: not by its maximal stars, but by some of its subtrees. Indeed, given k, the tangles of S_k live at the nodes[21] of the tree $\mathcal{T}_k$ for the tree set $T \cap S_k$. These trees $\mathcal{T}_k$ arise from $\mathcal{T}$ by contracting those of its edges that correspond to elements of $T \setminus S_k$. These trees $\mathcal{T}_k$ are nested in a way that reflects the nestedness of the tangles in $\vec{S}$ as subsets of $\vec{S}$. See [8, 12, 13] for more.

8.3 Tangle-precluding feature sets

The tree-of-tangles theorems from Section 8.2 offer a geometric picture of how the various tangles of S or in $\vec{S}$ are organized in a tree-like way. They produce a tree set $T \subseteq S$ which, if chosen minimal, partitions V into disjoint parts (possibly empty) that correspond exactly to the tangles of S or the maximal tangles in $\vec{S}$.

Our next theorem, the tangle–tree duality theorem, offers a similar tree set $T \subseteq S$ for the case that S has no tangles – more precisely, no $\mathcal{F}_n$-tangles for some n we have specified. Once more, T partitions V into disjoint parts, possibly empty. This time, however, the locations of T are not home to any tangle but the opposite: they lie in $\mathcal{F}_n$, and are thus too 'small' for an $\mathcal{F}_n$-tangle to live there.

Let us say that a tree set is *over* $\mathcal{F}$ if all its locations are elements of $\mathcal{F}$. Note that, while $\mathcal{F}$ may consist of arbitrary subsets of $\vec{S}$, those of its elements that occur as locations of a tree set will always be stars.

Theorem 3. (Tangle–tree duality theorem [12])
Let $\vec{S}$ be a submodular feature system, and let $\mathcal{F} = \mathcal{F}_n$ for some $n \geqslant 1$. Then exactly one of the following two statements holds:
  (i) *S has an $\mathcal{F}$-tangle;*
  (ii) *S contains a tree set over $\mathcal{F}$.*

See [12, 14, 15, 21] for more general versions of Theorem 3. These include versions for advanced feature systems as discussed in Section 7.6. The most general version, due to von Bergen [2], has an algorithmic proof that has been implemented in the software package published in [1].

It is easy to see why (i) and (ii) cannot happen at once, i.e., why the existence of a tree set over $\mathcal{F}$ precludes the existence of an $\mathcal{F}$-tangle. Indeed, suppose there are both a tangle τ as in (i) and a tree set T as in (ii). As $T \subseteq S$, the tangle τ induces a tangle of T. As T is over $\mathcal{F}$, the location of this tangle of T is an element of $\mathcal{F}$. Since it is also a subset of τ, this contradicts the fact that τ is an $\mathcal{F}$-tangle.

The deeper part of the theorem is that at least one of (i) and (ii) must always happen: that if S has no $\mathcal{F}$-tangle then it contains a tree set as in (ii), one which demonstrably precludes the existence of an $\mathcal{F}$-tangle. This is valuable, and it is the way in which Theorem 3 is usually applied. For example, suppose that some tangle-finding algorithm fails to find a tangle. This does not mean, yet, that no tangle exists. But if it can find, instead, a tree set as in (ii) of Theorem 3, we can check that indeed no tangle exists. The theorem thus assures us that whenever no tangle exists there is conclusive and easily checkable evidence for this.[22]

As a simple example for Theorem 3, let S consist of all the partitions of a set V, and let $\mathcal{F} = \mathcal{F}_1 \cup \{\{\{v\}\} : v \in V\}$. In other words, the only forbidden subsets in the $\mathcal{F}$-tangles we consider are inconsistent triples and singleton subsets $\{\{v\}\}$ of $\vec{S}$. We already noted in Section 2.4 that the only $\mathcal{F}_1$-tangles of S are the principal tangles τ_v focussed on their v. Since these contain the singleton features $\{v\}$, there are no $\mathcal{F}$-tangles of S for the above $\mathcal{F}$. If Theorem 3 holds for this $\mathcal{F}$, there should be a tree set $T \subseteq S$ over $\mathcal{F}$ that witnesses this.

While Theorem 3 does not formally apply to this $\mathcal{F}$, because it is not of the form $\mathcal{F}_n$, it does in fact hold for this $\mathcal{F}$ too. Indeed, our proof in Section 7.3 of the fact that every $\mathcal{F}_1$-tangle τ of S is of the form τ_v essentially produces a tree set as required in (ii). In that proof we constructed a decreasing sequence of subsets of V that ended with a singleton set $\{v\}$. Each set in the sequence was obtained from the previous one by dividing it into two subsets and choosing, as the next element for the sequence, the subset that was a feature in the given tangle τ. If we construct the same kind of recursive partition of V into singletons without focussing on any particular tangle τ, by dividing every non-singleton subset obtained into two smaller sets, we obtain an S-tree over our particular $\mathcal{F}$. Every local partition of a subset A into two smaller subsets B and C defines an inconsistent triple $\{A, \overline{B}, \overline{C}\}$ (complements taken in V), which will form a location of the tree set $T \subseteq S$ over $\mathcal{F}$ obtained in this way. Its other locations are the singleton sets $\{\{v\}\}$.

9

Order functions

In some of their most immediate applications, tangles are computed directly for a given feature system $\vec{S}$. For example, $\vec{S}$ might consist of the answers to a carefully designed questionnaire, received from the entire target population whose prevalent mindsets we wish to detect. The constitution of a representative body of delegates, as discussed in Section 5.4, is a typical example for such an immediate use of tangles.

However when S is larger, we may well wish to organize S into a hierarchy $S_1 \subseteq S_2 \subseteq \dots$ that begins with the most fundamental features while considering more specific ones later or not at all. We would then obtain a corresponding hierarchy of k-tangles, as discussed in Section 7.4. Our discussion there already indicated, however, that it may be best not to define an order function on S explicitly to set up this stratification, even when this might seem like a natural thing to do: there are issues of bias here, but also the technical issue that an explicitly defined stratification will lack some mathematical properties, such as submodularity, that we need for the more sophisticated tools of tangle theory.

In this chapter, therefore, we shall discuss ways of defining generic order functions: functions that are typically defined on *all* the partitions of a set V, and which do not appeal to any interpretation of the partition to which they assign an order. The order of a partition $\{A, B\}$ will usually be denoted by $|A, B|$, or briefly by $|s|$ if $s = \{A, B\}$.

As we shall see, these order functions, despite being generic, can take account of any hand-designed 'degrees of importance' we might wish to assign to some of these partitions, say those in our given set S, so that they assign low order to partitions that happen to lie in S and are deemed important. In other cases, such as in standard clustering applications for largely unknown datasets V, we have no pre-designed S anyway, and thus have to work with all the partitions of V simply for want of any detailed information.

Most of the generic order functions discussed in this chapter will be submodular. When they are, we shall not only be able to compute the tangles of the hierarchy $S_1 \subseteq S_2 \subseteq \ldots$ they define, but also to apply the more powerful tools of tangle theory such as our theorems from Chapter 8. On the whole, there is scope for some substantial and interesting mathematics in the question of how to choose order functions well, and how this choice influences how the k-tangles for different k interact.

Finally, S might be given to us by some other means, either directly inherent in our data or produced by some existing clustering method. For example, if V is the set of pixels of an image, then S arises naturally as the set of lines through this image that cut it in two. Or S might be produced by a neural network trained on known data – say, one that has already learned to distinguish chairs from tables. In Chapter 10 we shall look at some mathematical ways of extracting a suitable starting set S of partitions of V from our data. To all these S we can then apply the order-based hierarchies to be discussed in this chapter.

We begin in Section 9.1 by designing an order function on all the partitions of V that is based on a given measure of similarity between its elements. If V is a set of texts, for example, then two of these might be deemed similar if they use similar words or syntax, something we can easily measure. The idea, then, is to assign high order to partitions that divide many similar points, thereby penalizing such partitions and making it harder for them to be adopted into S_k for small k. Generally, similarity measures will depend heavily on the context in which we seek to apply tangles. However, the way in which they translate into order functions follows some general principles, which we discuss in Section 9.1.

In our questionnaire scenario, two individuals in V might be deemed 'similar' if they answered many of the questions in the same way. In the dual system, two questions would be similar if many $v \in V$ answered them in the same way. Importantly, if we define similarity between individuals in V in terms of how they answered the questions in a questionnaire Q, we obtain an order function on the set S of all the partitions of V. The questions in Q are themselves such partitions, so in particular we obtain an order function on Q. This is perfectly fine. Essentially, a question will receive large order, and thus be deemed less significant, if it divides the population very differently from how the other questions divide it, while questions receive low order, and are deemed important, if they divide the population in a way that is typical for many of the other questions.

In Section 9.2 we extend this idea to order functions on advanced feature systems of $V \to \mathbb{R}$ functions, as discussed in Section 7.6. The idea is to give high order to functions f that assign to many similar elements of V very different values. For example, two elements u, v of V might be deemed 'similar' if $|\vec{s}(u) - \vec{s}(v)|$ is small for many advanced features $\vec{s}$. In the dual system, two advanced features would be deemed similar if they assign similar values to many elements of V. As before, such an order function will be defined on all $V \to \mathbb{R}$ functions, even though this definition may refer to some particular functions which, for example, stem from a questionnaire.

When we define the similarity between elements of V in terms of a questionnaire Q in this way, arbitrary partitions of V will receive low order if they divide V similarly to how most or many of the partitions in Q do: if they are 'like' many elements of Q in this sense. This 'likeness' between partitions can be measured directly, giving rise to another way of defining order functions on the set S of all partitions of V. Similarly, the order of arbitrary $V \to \mathbb{R}$ functions based as above on how much these functions resemble some given functions (e.g., those coming from a questionnaire) can be defined directly by defining a degree of 'likeness' between $V \to \mathbb{R}$ functions. We shall do this in Section 9.3.

The order functions discussed in Section 9.4 take a different tack, in that they measure the similarity of either points or features in a more qualitative rather than quantitative way. The idea is that if the sides of a partition of V are structurally simple, because the way in which they are divided by potential features conforms to some simple patterns, we might assign that partition low order. Similarly if the point similarities across the partition are structurally simple, e.g. because many points on one side are similar to some common set of points on the other side, we might assign that partition low order. Both these can happen even when our partition divides many similar pairs.

The order function introduced in Section 9.5 combines these two approaches from Section 9.4 by assessing not only the structure on each side of a partition, or the structural interaction between the two sides, but by comparing the internal structures of the two sides with each other.

In technical but suggestive terms, these order functions reflect the *mutual information* that the two sides of a partition hold about the other side, a concept from information theory based on something called 'discrete entropy'. These order functions assign high order to partitions whose sides are structured similarly by some potential features we consider important – for example, by how their elements answered the questions in a questionnaire Q. In the example, a partition receives high order – and is therefore discouraged – if either the people on the two sides answered many questions in the same way, or if they answered many questions in opposite ways. In either case the structure of one side, as imposed by the partitions in Q, tells us a lot about the structure of the other side, so each contains a lot of 'information' about the other.

In Section 9.6 we apply this idea more generally to devise entropy-based order functions for the dual of a given feature system, as defined in Section 7.7. While in Section 9.5 we used entropy to assess how a partition of V increases the structure, or 'information', inherent in how the elements of its two sides answered a given questionnaire, we now use it to define an order on the partitions of the set of questions rather than the people. The idea is to determine how much the answers, given by all the people in V, to the questions on one side of such a partition tell us about their answers to the questions on the other side.

In Section 9.7 we apply these entropy-based order functions for the dual system to the primal system as envisaged in the first three sections of this chapter. The functions we considered there defined the order of a given partition s of V, or of an advanced feature $\vec{s}: V \to \mathbb{R}$, by summing certain penalties $\sigma(u, v)$ or $\langle u, v \rangle$ over all pairs $u, v \in V$ that were divided by s or assigned different values by $\vec{s}$. These sums were taken over all the questions or potential features $q \in Q$ giving rise to that system, making the penalty increase with every q on which u and v differed.

The new idea now is to base this sum not on the questions in Q themselves but on typical, if hypothetical, questions that summarize those in Q: the tangles of the dual feature system. In a questionnaire scenario, these tangles might represent topics of the questions in Q, or social groups in the population V that determine how they answered the questions in Q.

In Section 9.8 we discuss some ways of tweaking order functions so as to encourage balanced partitions, by reducing their order. Strongly unbalanced partitions, those that split off just one or two elements of V, carry no useful information, because all interesting tangles will specify them in the same way, orienting them towards their larger side.

In Section 9.9, finally, we discuss some further aspects of order functions that are not specific to any particular one of them, and are therefore best discussed together. One of these is how to weed out 'irrelevant' partitions, partitions that reflect information entirely unrelated to our object of study. Such partitions can be identified mechanically, without any recourse to interpretation. Most of our order functions assign these high order, which excludes them from the sets S_k whose tangles we compute. Conversely, we shall look at how we can tweak a generic order function so as to assign low order to some features we deem to be important, without losing the property of being generic and, as a result, submodular.

When we compare the various order functions discussed in this chapter, it is important to bear in mind that no effort has been made to make them comparable in absolute terms, by their numerical values. If such comparison is desired, the two functions to be compared first have to be scaled appropriately. Similarly, if we wish to compare the orders of partitions of sets of different size, such as of the entire ground set V and a sampled subset $U \subseteq V$, we may have to normalize our order functions first.[1] This has not been done in how they are presented here, so as not to clutter their appearance with normalization terms irrelevant to our discussion of how these order functions work.

9.1 Order functions based on point similarity

In this section we look at some generic ways of defining order functions on feature systems $\vec{S}$ that make no reference to any interpretation of a feature whose order is to be decided. However, these order functions do reflect the structural properties of the features in $\vec{S}$: the order assigned to a given $\vec{s} \in \vec{S}$ will be based on how the way it partitions V compares with how the other features in $\vec{S}$ do that.

But then, just as this overall structure of $\vec{S}$ on V determines the order of a particular partition $s \in S$, it can determine the order of an arbitrary partition of V. Hence our order functions will be defined not just on S but on all the partitions of V. Such order functions, and the hierarchy $S_1 \subseteq S_2 \subseteq \ldots$ associated with them, can then be used either on the given S, or on the set of all the partitions of V, or on some subset of these computed by other methods.

However to better understand our task of assigning an order to an arbitrary partition $\{A, B\}$ of V, let us begin with the cases where we do have an intuitive aim for how to do this: where the potential features in S come with an interpretation, and $\{A, B\}$ happens to lie in S. In that case we said that $|A, B|$ should be small if its complementary features A and B were 'basic', or 'fundamental'. Let us try now to mimic this in purely structural terms: without appealing to that interpretation, but allowing ourselves to refer to how the elements of S – especially those other than $\{A, B\}$ – partition V.

One aspect of $s = \{A, B\}$ being 'fundamental' in any interpretation is that it divides V along natural lines of dissimilarity: that it does not divide many otherwise similar objects.[2] For example, the feature of having legs is fundamental in this sense for furniture: most chairs have legs, most tables have legs, and few beds or wardrobes have legs. So the feature of having legs does not divide many pairs of objects that are otherwise similar in that both are chairs, both are tables, both are beds or both are wardrobes. The feature of being made almost entirely of wood, by contrast, is not fundamental in this sense: there are about as many chairs and tables that have this feature as there are chairs and tables that do not have it, so it divides many pairs of chairs and many pairs of tables.

This aspect of being fundamental can be expressed in an abstract way – with reference to S, but without interpretation – as soon as being 'similar' can be. But similarity with respect to the features in $\vec{S}$ can often be expressed in a formal way too. For example, it may be appropriate to think of two objects in V as similar if they share many of the features in $\vec{S}$: if, for many $s \in S$, they both specify s as $\vec{s}$ or they both specify it as $\overleftarrow{s}$.[3] Let us define such similarities more formally.

Let $V^{(2)}$ denote the set of all the 2-element subsets $\{u, v\}$ of V, our set of objects. Note that only sets $\{u, v\}$ with $u \neq v$ lie in $V^{(2)}$. Let us call any function $\sigma \colon V^{(2)} \to \mathbb{R}$ a *similarity measure* on V. To simplify notation, we usually write $\sigma(u, v)$ instead of $\sigma(\{u, v\})$; then $\sigma(u, v) = \sigma(v, u)$, since $\{u, v\}$ and $\{v, u\}$ denote the same 2-element set. Similarity measures can thus be viewed as symmetric functions σ on the pairs of distinct elements of V, with $\sigma(v, v)$ undefined for all $v \in V$.

Our earlier idea now was to let $\sigma(u, v)$ count the features that u and v have in common, setting

$$\sigma(u, v) := \left| \{\, s \in S : u(s) = v(s) \,\} \right|$$

where

$$v(s) := \begin{cases} \vec{s} & \text{if } v \text{ specifies } s \text{ as } \vec{s} \\ \overleftarrow{s} & \text{if } v \text{ specifies } s \text{ as } \overleftarrow{s} \end{cases}$$

and $u(s)$ is defined analogously. Thus, u and v have large *similarity* $\sigma(u, v)$ if they share many of the features in $\vec{S}$. Since σ makes no difference between the two specifications $\vec{s}$ and $\overleftarrow{s}$ of a given $s \in S$, we shall call it *unbiased*.

Often, however, the potential features $s \in S$ have default specifications $\vec{s}$ that are more distinctive than their inverses $\overleftarrow{s}$. In our furniture example, $\vec{s}$ might denote the feature of being collapsible, which is more distinctive than not being collapsible. In this case we might want to count only these distinctive features in the definition of σ. The alternative *biased* definition

$$\sigma(u, v) := \left| \left\{ s \in S : u(s) = v(s) = \vec{s} \right\} \right|$$

reflects the fact that we consider two collapsible pieces of furniture as more like one another than two arbitrary non-collapsible pieces of furniture.

Based on any given similarity measure σ on V, we can define an order function on the partitions of V that assigns low order to partitions that do not split many pairs of similar objects:

$$|A, B| := \sum_{a \in A} \sum_{b \in B} \sigma(a, b). \tag{O1}$$

Note that, as desired, $|A, B|$ is large if there are *many* pairs (a, b) with *high* similarity $\sigma(a, b)$, and small otherwise.

In principle, the similarity measure σ in (O1) is allowed to take negative values – as, in fact, are orders of partitions in general (although this rarely happens). If σ takes non-negative values only, as in our two examples, the order function (O1) is submodular.[4]

When we compute the order of a partition $\{A, B\}$ in this way we take, for every $s \in S$ referred to in the definition of σ, a sum over all pairs (a, b) with $a \in A$ and $b \in B$. Unless our partition divides V very unevenly, the order of magnitude of this sum is quadratic in $n = |V|$: give or take a multiplicative constant, there are about n^2 such pairs (a, b) that determine the value of $|A, B|$.

Let us now define an order function, still based on either of our two similarity measures σ, for which the number of summands for each reference $s \in S$ is only linear in n. This difference can have a profound computational impact, and illustrates the degree of influence we can wield through our choice of an order function on which our hierarchy $S_1 \subseteq S_2 \subseteq \ldots$, and hence the tangles to be found, depends.

Let us begin by rewriting our earlier unbiased order function as

$$|A, B| = \sum_{a \in A} \sum_{b \in B} \sigma(a, b) = \sum_{\vec{s} \in \vec{S}} |A \cap \vec{s}| \cdot |B \cap \vec{s}|$$

and the biased one as

$$|A, B| = \sum_{a \in A} \sum_{b \in B} \sigma(a, b) = \sum_{s \in S} |A \cap \vec{s}| \cdot |B \cap \vec{s}|.$$

We now replace the quadratic expression of $|A \cap \vec{s}| \cdot |B \cap \vec{s}|$ under the sum on the right with a term that is linear in n, letting

$$|A, B| := \sum_{\vec{s} \in \vec{S}} \min \left(|A \cap \vec{s}|, |B \cap \vec{s}| \right) \tag{O2}$$

for the unbiased version and

$$|A, B| := \sum_{s \in S} \min \left(|A \cap \vec{s}|, |B \cap \vec{s}| \right) \tag{O3}$$

for the biased version. These order functions, too, are submodular [11].

Our new 'linear' order functions are qualitatively similar to our old 'quadratic' ones: they still assign the largest values to partitions that split many of the features $\vec{s}$ we care about, and which therefore feature in the sum, about equally. But for features $\vec{s}$ that $\{A, B\}$ splits about equally we now assign a penalty of at most $|\vec{s}|/2$, while our earlier order functions assigned penalties of up to $(|\vec{s}|/2)^2$. This contrasts with the difference in penalties for features $\vec{s}$ that $\{A, B\}$ splits more unevenly: if A has only one element in $\vec{s}$ and B contains all the others, we assign a penalty of 1 in our new order but one of $|\vec{s}| - 1$ in the old, an increase by a factor of nearly $|\vec{s}|$ rather than $|\vec{s}|/2$.

Thus, all our order functions so far 'prefer' partitions that do not split many of the larger $\vec{s}$ evenly to those that do, but the new linear order functions do so only half as strongly as the old quadratic ones.

To conclude this section, let us introduce another type of similarity-based order function, one that not only discourages the splitting of similar pairs of elements of V but, in addition, actively encourages keeping similar pairs on the same side. Although not technically necessary, let us continue to assume that σ takes non-negative values only; this will make it easier to find intuitive informal language to describe the new order function.

An encouragement to keep similar pairs of points on the same side can be built into all four of the order functions we have discussed so far; we illustrate the idea on (O1). Let us modify this to

$$|A, B| := \sum_{a \in A} \sum_{b \in B} \sigma(a,b) \; - \sum_{\{a,a'\} \in A^{(2)}} \sigma(a,a') \; - \sum_{\{b,b'\} \in B^{(2)}} \sigma(b,b') \,. \qquad \text{(O4)}$$

Thus, for every pair a, b of points it splits, the partition $\{A, B\}$ accrues a penalty of $\sigma(a,b)$, while for every pair a, a' or b, b' not split it is rewarded by a 'negative penalty' $\sigma(a,a')$ resp. $\sigma(b,b')$. As before, these penalties are the stronger, positive or negative, the more similar those two points are. The order function (O4) is submodular if all the $\sigma(u,v)$ are non-negative.

Such additional encouragement to keep similar points together is technically redundant. But we shall see in Section 9.8 that it can easily be tweaked, while keeping it submodular, to make low-order partitions more balanced – an issue that has practical implications for how fast tangles can be computed.

Although similarity-based order functions can be defined on all partitions of V, they can equally well be used when we quite deliberately restrict our attention to the original partitions of V from S. We might do this, for example, when our features come with interpretations on which we would like to base similar interpretations of the tangles we find.

9.2 Advanced order functions based on point similarity

In this section we indicate how similarity-based order functions can be defined on real-valued functions $f\colon V \to \mathbb{R}$, the elements of the 'advanced' feature systems we introduced in Section 7.6.

When these f take values 1 or -1 only[5] we can interpret them as oriented partitions of V, which remains the most important case also of advanced features. And even the order functions we shall define below will be then the same, up to a constant factor, as (O1) in Section 9.1. But while rewriting partitions as functions may lessen some of their intuitive immediacy, it also comes with a benefit: it will enable us to compare the order functions from Section 9.1, rewritten in a more algebraic way, with order functions that are innately algebraic. We shall consider such algebraic order functions in Sections 9.3 and 9.4.

Since the two specifications of a potential feature should have the same order, we must take care that when we define our order functions on the features f directly we ensure that $|-f| = |f|$ for all f. All the functions $f \mapsto |f|$ we consider here will satisfy this, so we shall not normally mention this requirement again.

We shall first suggest a way of defining an order function $f \mapsto |f|$ on the basis of *any* given similarity measure on the pairs of our objects $v \in V$; for now, this can be any symmetric real-valued function on the pairs of distinct elements of V, or equivalently any real-valued function on the set $V^{(2)}$ of the 2-element subsets of V. In a second step we shall then suggest a particular way of defining such a similarity measure in our usual questionnaire context, with a set Q of questions that can take answers between -5 and 5, say, so that the similarity of $u, v \in V$ will be determined by how similarly u and v answer the questions in Q.

So let us first assume that a similarity measure $\sigma \colon V^{(2)} \to \mathbb{R}$ is given. For now, σ may take negative values.

A natural way to define the order $|f|$ of a feature $f \colon V \to \mathbb{R}$ is to let

$$|f| := \sum_{\{u,v\} \in V^{(2)}} \sigma(u, v) \cdot \big(f(v) - f(u) \big)^2. \tag{O5}$$

This function is submodular if all the $\sigma(u, v)$ are non-negative.

In the above sum, the expression $(f(v) - f(u))^2$ is just one of many possible ways of measuring how far apart the f-values of u and v are: we could have used $|f(v) - f(u)|$ instead. A reason for choosing the quadratic expression above is that this makes it a little easier to compute $|f|$, and to compare it with (O6) below.

So why might this be a sensible order function for advanced features? In our traditional binary setup we aimed to assign a partition of V low order if it did not separate many similar points in V; see (O1). Our order function (O5) does the same for advanced features: for every pair u, v of points to which f assigns different values we assign f a penalty of $\sigma(u, v)$, just as we did in (O1). But now this penalty is weighted, in addition, with *how* different the f-values of u and v are: the greater the difference between $f(u)$ and $f(v)$, the more will the similarity $\sigma(u, v)$ of u and v contribute to the overall penalty on f, i.e., to its order.

Another difference to the concrete choices for σ in Section 9.1 is that, for now, σ may take negative values. If $\sigma(u, v) < 0$, then assigning different values to u and v lowers the order of f. When f encodes a partition of V in that it takes values -1 and 1 only, this means that f is not only penalized for splitting similar pairs u, v (for which $\sigma(u, v) > 0$) but is also rewarded for splitting dissimilar pairs u, v (for which $\sigma(u, v) < 0$). As we shall see in Section 9.8, this has consequences for how balanced the partitions of lowest order are.

Let us now adapt to our more general setting the order function (O4) discussed last in Section 9.1, the one that rewards partitions for *not* splitting similar pairs. As there, let us assume when we discuss this informally that σ takes non-negative values only, although this is not technically necessary at this point.[6] Given $f: V \to \mathbb{R}$ as before, consider

$$|f| := -\sum_{\{u,v\} \in V^{(2)}} f(u)\, \sigma(u, v) f(v)\,. \tag{O6}$$

When f takes values -1 and 1 only, and we rename $f^{-1}(-1)$ as A and $f^{-1}(1)$ as B, then this $|f|$ coincides with (O4): the contribution to the order of $\{A, B\}$ for any pair u, v it splits is $\sigma(u, v) \geqslant 0$, since $f(u)f(v) = -1$ for such u and v and $-(-1)\sigma(u, v) = \sigma(u, v)$, whereas for pairs u, v on the same side of the partition, either both in A or both in B, the contribution is $-\sigma(u, v)$.

For general $f: V \to \mathbb{R}$ things are similar: the order of f as in (O6) is increased for every pair u, v on which f takes values with different signs ($-$ or $+$), so assigning qualitatively different values to similar pairs is penalized, while for pairs u, v on which f takes values with the same sign the order is reduced, so assigning qualitatively similar values to similar pairs is rewarded.

Let us now take a look at what might be a sensible similarity measure for advanced feature systems in a questionnaire scenario. Let Q be the set of questions in a survey answered by the individuals in V, where each question could have answers in a range between 'disagree entirely' and 'agree entirely'. In our model, Q is a set of functions $q: V \to \mathbb{R}$ with integer values between -5 and 5.[7]

How can we define a similarity measure on V based on how the people in V answered the questions in Q? One way, analogous to our considerations above, is to define

$$\sigma(u,v) := \sum_{q \in Q} \left(C - (q(v) - q(u))^2 \right),$$

for distinct $u, v \in V$, where $C \in \mathbb{N}$ is a constant we may choose. If we choose C large enough to make all the values of σ non-negative, then this will make (O5) and (O6) submodular. If we choose it smaller, or omit it altogether, then low-order partitions $f: V \to \{-1, 1\}$ for (O5) will be more balanced. Those for (O6) are already more balanced; see Section 9.8.

In Section 9.1 our first approach was to have $\sigma(u, v)$ count the number of questions to which u and v give the same answer. If these answers take values 0 or 1,[8] then this count equals the sum, over all questions q, of the values $C - (q(v) - q(u))^2$ with $C = 1$, since this value is 1 for questions q on which u and v agree, while it is 0 when they disagree.

Our definition of $\sigma(u, v)$ above extends this to questions with a larger numerical range of answers, taking into consideration also how strongly u and v disagree on a question q when they do: the larger this difference, the number $(q(v) - q(u))^2$, the lower the similarity between u and v regarding q. As earlier, the term $(q(v) - q(u))^2$ is only one way of many to measure how much u and v disagree on q: we could replace it with $|q(v) - q(u)|$, or with whatever distance function seems to fit the intended application best.

As in a binary context there is ample scope here for modifications depending on the context in which we intend to measure similarities. Just as we modified the definition of σ in Section 9.1 to take into account only positive answers (when this seemed more appropriate), we could now choose to limit the sum to only those q that are positive on both u or v, or on at least one of them, or for which $q(u)$ and $q(v)$ have the same sign (are both positive or both negative), and so on.

To conclude this section let us rewrite our two order functions (O5) and (O6) in matrix form. This will help us compare them with each other, seen from a new angle. *For the remainder of this section, we assume that σ takes non-negative values only.*

We start with rewriting

$$|f| := -\sum_{\{u,v\} \in V^{(2)}} f(u)\,\sigma(u,v)f(v). \tag{O6}$$

Let K be the symmetric $n \times n$ matrix indexed by the elements of V with off-diagonal entries $-\sigma(u, v)$ in row u and column v, and diagonal entries

$$\sigma(u, u) := 0$$

to ensure everything is defined. If we think of a function $f: V \to \mathbb{R}$ as the $n \times 1$ matrix, indexed by V, that has $f(v)$ in position v, then (O6) translates to

$$|f| = -\tfrac{1}{2} \sum_{(u,v) \in V^2} f(u)\, \sigma(u, v) f(v) = \tfrac{1}{2} f^{\top} K f. \tag{O6'}$$

The factor of $1/2$ comes from the fact that we are now summing over the pairs (u, v), where (u, v) and (v, u) give rise to separate equal summands, rather than over the 2-sets $\{u, v\}$ where this summand featured only once.

Rewriting the order function

$$|f| := \sum_{\{u,v\} \in V^{(2)}} \sigma(u, v) \cdot \big(f(v) - f(u)\big)^2 \tag{O5}$$

is a little more complicated. To get started, let us fix a linear ordering on V. Next, we encode the similarities $\sigma(u, v)$ on V in a matrix W, as follows. The $n = |V|$ rows of W are indexed by the elements of V, its $m = \binom{n}{2}$ columns by the elements of $V^{(2)}$, the unordered pairs $\{u, v\}$ of distinct elements of V. Now let column $\{u, v\}$ have $-\sqrt{\sigma(u, v)}$ in row u and $\sqrt{\sigma(u, v)}$ in row v if $u < v$ in our linear ordering on V; in all other rows this column has zeros.

As before, we think of a function $f: V \to \mathbb{R}$ as the $n \times 1$ matrix, indexed by V, that has $f(v)$ in position v. Then $W^{\top} f$ is the $m \times 1$ matrix that has

$$\sqrt{\sigma(u, v)} \cdot \big(f(v) - f(u)\big)$$

in position $\{u, v\}$ (again with $u < v$). Let

$$L := WW^{\top}.$$

This is a symmetric $n \times n$ matrix whose entries, unlike in the case of W, no longer depend on the linear ordering we imposed on V: they just depend on the pair (u, v) marking the position of that entry (see below).

This matrix, moreover, enables us to write our order function (O5) compactly as follows:

$$f^\mathsf{T} Lf = f^\mathsf{T} WW^\mathsf{T} f = (W^\mathsf{T} f)^\mathsf{T} (W^\mathsf{T} f) = \sum_{\{u,v\} \in V^{(2)}} \sigma(u,v) \cdot \big(f(v) - f(u)\big)^2,$$

so

$$|f| = f^\mathsf{T} Lf \quad (\geqslant 0). \tag{O5$'$}$$

The matrix L is known as a *Laplacian*. By definition, $L = WW^\mathsf{T}$ has off-diagonal entries $-\sigma(u,v)$, just like K. Its diagonal entries, however, are $\sum_v \sigma(u,v)$ in position u, the sum being over all $v \neq u$. This has the effect that $L\mathbb{1} = 0 \in \mathbb{R}^n$, where $\mathbb{1} = (1, \ldots, 1)^\mathsf{T}$. If we define

$$\sigma(u,u) := -\sum_{v \neq u} \sigma(u,v)$$

then (O5$'$), and hence (O5), can be written similarly to (O6$'$), as

$$|f| = f^\mathsf{T} Lf = -\sum_{(u,v) \in V^2} f(u)\, \sigma(u,v) f(v). \tag{O5$''$}$$

(But remember that $\sigma(u,u)$ is defined differently in (O5$''$) and (O6$'$).)

We shall meet L again in Section 10.2, but for the very different purpose of generating S itself in certain non-questionnaire contexts.

Expressing our order functions $f \mapsto |f|$ from (O5) and (O6) in terms of L and K can be used conveniently to show that they are submodular in the realm of real-valued functions on V [25]. This fact is independent of our choice of σ, as long as σ takes non-negative values only.

9.3 Order functions based on feature similarity

In Section 9.1 we looked at ways of assessing whether or not $s = \{A, B\}$ is a good way of splitting V by looking at the pairs $\{a, b\}$ with $a \in A$ and $b \in B$: if our data suggests that many of these pairs are 'similar', then s would be considered as a bad way to split V, so we would assign it high order to exclude it from our feature systems $\vec{S}_k$ for small k. Similarly, we would assign an advanced feature $f\colon V \to \mathbb{R}$ high order in Section 9.2 if it gave many similar points very different values.

In a questionnaire context, a natural way to define these similarities $\sigma(u,v)$ was to measure how similarly u and v answered the questions.

But then, would it not be more natural to compare those partitions of V, or the $V \to \mathbb{R}$ functions whose order we wished to define, directly with those questions? After all, these partitions or functions are of the same 'kind' as those questions: both are partitions of V, or $V \to \mathbb{R}$ functions. Such a direct comparison, it seems, might allow us to take more information into account than what can be compressed into the values $\sigma(u, v)$ of just pairs of points, even if these are defined in terms of how u and v answered Q.

But how exactly should a partition s or function f compare with Q in order to be assigned high or low order? The idea we discussed in Section 9.1 was to 'prefer' partitions, by assigning them low order, that crossed the partitions defined by Q as little as possible: partitions that split few pairs u, v not split by many $q \in Q$. In other words: we assigned low order to the partitions that were most 'like' the given partitions in Q. A more qualitative version of this in the context of advanced features would be to prefer functions $f: V \to \mathbb{R}$, by assigning them low order, that are 'like' many of the functions $q: V \to \mathbb{R}$ in Q. So let us try to implement this idea.

Since our plan is to define an order $|f|$ directly on features $f: V \to \mathbb{R}$ now, rather than on potential features $\{f, -f\}$, we have to make one adjustment. Our order function should not prefer functions f that are 'like' many of the functions $q \in Q$ in the sense of taking similar values; rather, it should prefer functions f that are strongly *correlated* to many $q \in Q$, positively or negatively, in the sense that they either take very similar or nearly opposite values as those q. Indeed, we have to require that $|f| = |-f|$, but the two functions f and $-f$ take exactly opposite values and hence cannot both be 'like' many $q \in Q$ in the sense of taking similar values as those q do.

A standard way of measuring the correlation of two functions is what statisticians call their *covariance*. To keep things simple, let us instead work with a similar but better known measure for the correlation of two real n-vectors: their 'dot product', or canonical inner product. Formally, the *dot product* of two vectors $x = (x_1, \ldots, x_n)$ and $y = (y_1, \ldots, y_n)$ is the real number $\langle x, y \rangle = \sum_{i=1}^{n} x_i y_i$.

We can apply this to measure the correlation of two $V \to \mathbb{R}$ functions f and g in the same way, letting

$$\langle f, g \rangle := \sum_{v \in V} f(v) \cdot g(v).$$

If our questions q take values in $\{-5,\ldots,5\}$, for example, and a feature $f\colon V \to \mathbb{R}$ is positively correlated to a question $q \in Q$ in the informal sense that many of the people $v \in V$ with $f(v) > 0$ answer q positively, and many of those with $f(v) < 0$ answer q negatively, then $f(v)\cdot q(v) > 0$ for all these v since $f(v)$ and $q(v)$ have the same sign, which makes $\langle f,q\rangle$ positive. Similarly, if f is negatively correlated to q in the informal sense that f and q take values with different signs on many v, then those $f(v)\cdot q(v)$, and hence $\langle f,q\rangle$, will be negative.

Now to encourage the choice of f, by making $|f|$ small, if f is strongly correlated to many $q \in Q$ (positively or negatively), we can define

$$|f| := -\sum_{q\in Q}\langle f,q\rangle^2 \quad (\leqslant 0). \tag{O7}$$

If desired, we can make these values positive by adding some large constant; this is immaterial to what follows.

Following the example of Section 9.2, let us rewrite our new order function in matrix form; this will help us compare it with those earlier ones. Let M be the $n \times m$ matrix whose $n = |V|$ rows are indexed by the elements of V, whose $m = |Q|$ columns are indexed by the elements of Q, and which has entries $q(v)$ in row v and column q. Note that this matrix encodes the entire interaction between V and Q: the interaction on which we are aiming to base our order function $f \mapsto |f|$.

The terms $\langle f,q\rangle$ on which our new order function is based are recorded, separately for all the $q \in Q$, in $M^\top f$: this is an m-vector, indexed by the elements of Q, whose entry in position q is precisely

$$\sum_{v\in V} f(v)\cdot q(v) = \langle f,q\rangle.$$

Our new $|f| = -\sum_{q\in Q}\langle f,q\rangle^2$, therefore, is simply the negative of the dot product of this m-vector $M^\top f$ with itself:

$$|f| = -(M^\top f)^\top (M^\top f) = -f^\top M M^\top f.$$

Hence for $J := -MM^\top$ we obtain our desired expression of $|f|$ in matrix form:

$$|f| = f^\top J f. \tag{O7'}$$

Let us take a closer look at this matrix J. Its negative, $-J = MM^\top$, is an $n \times n$ matrix indexed by the elements of V. And its entries are beautifully simple: in row u and column v this matrix $-J$ has the dot product

$$\langle u, v \rangle = \sum_{q \in Q} q(u) \cdot q(v)$$

of u and v, viewed as $Q \to \mathbb{R}$ functions $q \mapsto q(u)$ and $q \mapsto q(v)$. Spelling out our compact representation of $|f|$ as $f^\top J f$ explicitly, we obtain

$$|f| = -\sum_{(u,v) \in V^2} f(u) \langle u, v \rangle f(v), \qquad (O7'')$$

which looks a bit like (O6$'$) and (O5$''$), with $\langle u, v \rangle$ replacing $\sigma(u, v)$. What does this tell us about $|f|$?

The terms $\langle u, v \rangle$ measure the correlation of u and v viewed as $Q \to \mathbb{R}$ functions, $q \mapsto q(u)$ and $q \mapsto q(v)$.[9] They are positive and large if u and v agree strongly on many of the questions $q \in Q$, and negative with large absolute value if they disagree strongly about many q. In short, $\langle u, v \rangle$ once more measures the like-mindedness of u and v regarding the questions in Q.

The most obvious difference between $\langle u, v \rangle$ and the similarity measure

$$\sigma(u, v) = \sum_{q \in Q} \left(C - (q(v) - q(u))^2 \right)$$

on which (O6$'$) and (O5$''$) were based is that in $\langle u, v \rangle$ the similarity between $q(u)$ and $q(v)$ for fixed q is now measured multiplicatively, as $q(u) \cdot q(v)$, rather than additively in terms of $q(v) - q(u)$.

Another difference is that we had $\sigma(u, u) = 0$ in (O6$'$), which neutralizes the effect of $f(u)$ in the term for (u, u). By contrast, the term $\langle u, u \rangle$ in (O7$''$) encourages functions f that take large values on elements u with strong views, by reducing their order, while the diagonal entries of $\sigma(u, u) = -\sum_v \sigma(u, v)$ in (O5$''$) discourage them unless the other $v \in V$ hold similar strong views.[10]

Finally, $\langle u, v \rangle$ can always take negative values, whereas $\sigma(u, v)$ does so only if the constant C is small enough, such as $C = 0$. This has implications for submodularity, but also, when f encodes a partition of V by taking values in $\{-1, 1\}$ only, on how balanced these partitions are. More on this in Section 9.8.

Although these differences may be seen as improvements over our earlier σ-based order functions, the overall similarity between the two ways of defining $|f|$ may come as a surprise. We set out with high aims to base $|f|$ more directly on the interaction between V and Q, rather than just on its quantitative aspects expressed in the similarities $\sigma(u,v)$ of pairs. And we defined $|f|$ naturally to achieve this; indeed we found a simple way to express it directly in terms of the matrix M that encodes this entire interaction. But now we find that only superficial aspects appear to have changed – except that L is only defined when σ takes non-negative values only, whereas J and K are always defined.

However, this can also be viewed as an unexpected positive insight: the similarities $\sigma(u,v)$, when taken over *all* the pairs $\{u,v\}$ of points in V, appear to encode more of the overall interaction between V and Q than meets the eye. As a consequence, the order functions $f \mapsto f^\top K f$ from (O6′) and $f \mapsto f^\top L f$ from (O5′) based on Q-based similarities $\sigma(u,v)$ remain serious contenders for $|f|$, even compared with our more sophisticated new $f \mapsto f^\top J f$, when all these similarities are non-negative.

Our new order function $f \mapsto |f|$ from (O7′) is submodular if and only if all the $\langle u,v \rangle$ are non-negative. We could achieve this by letting the $q \in Q$ take non-negative values only. But this has consequences for the interpretation of $\langle u,v \rangle$. For example, if Q models yes/no questions by letting the q take values 1 and -1, then $\langle u,v \rangle$ counts the number of questions on which u and v agree, minus those on which they disagree; if Q models the same yes/no questions by letting the q take values 1 and 0, then $\langle u,v \rangle$ counts the number of questions which u and v both answer as yes. There are applications where such bias is welcome, as in our example of biased order functions in Section 9.1. But in other applications it may not be what we want.

9.4 Algebraic order functions

In the realm of our standard binary feature systems we seek to identify the most natural ways of splitting a set V in two, to find its 'bottlenecks', and then to assign these some low order so that our tangles can orient those bottlenecks successively with increasing order.

Where these bottlenecks are is not determined by V itself as a set, but depends on the structure, or data, on V. When we defined our order functions in Section 9.1, we focussed our view of this data on just the pairs in V, assigning a partition $\{A, B\}$ of V low order if it split few pairs deemed similar in terms of our data. Our approach in this section and the next is to look more globally at the internal structure of the data on each of the two sides A and B, or at the interaction of the data across the split.

For example, suppose the data on V is organized in the form of a vector space, of which the data on A and on B alone forms subspaces. Then we can measure how much these subspaces interact by computing their dimensions: if the dimensions of the two subspaces add up to no more than the dimension of the whole space, then these subspaces do not interact at all, and $\{A, B\}$ would be a particularly natural split of V.[11]

For simplicity, let us assume in this section that our data on V is given as in the standard questionnaire setting: in the form of a set Q of functions $q\colon V \to \mathbb{R}$, with integer values in $\{-5, \ldots, 5\}$ say, which we think of as questions answered by all the $v \in V$. We can equally think of the elements of V as functions $v\colon Q \to \mathbb{R}$, also with values in $\{-5, \ldots, 5\}$, that map every question q to its answer $v(q) = q(v)$ given by v. Generalizations to other settings, or to advanced features instead of partitions, are possible but not necessary in order to explain the essential ideas that are new in this section.

Let us start with a particularly simple example of defining the structure 'across' a partition $s = \{A, B\}$ of V more globally than in Section 9.1. There, we defined its order $|s|$ as the sum of similarities $\sigma(a, b)$ with $a \in A$ and $b \in B$. Let us simplify the example further and assume that these similarities are either 1 or 0; then $|s|$ is simply the number of similar pairs a, b across the split.

Rather than just counting the number of such pairs, let us now look more closely at where they are. To facilitate this, let N denote the $n_A \times n_B$ matrix whose rows are indexed by the n_A elements of A, whose columns are indexed by the n_B elements of B, and which has 1 in position (a, b) if a and b are deemed similar (based on their answers to Q or not), and 0 otherwise.[12] The columns of N, then, are 0/1-vectors of length n_A, which we can view as elements of the n_A-dimensional vector space over the 2-element field $\mathbb{F}_2 = \{0, 1\}$.[13] In this vector space $\{0, 1\}^A$ we can now look at how large a subspace the columns of N span, and define $|s|$ as the dimension of this subspace, the *rank* of the matrix N:[14]

$$|A, B| := \operatorname{rank} N.$$

Linear algebra tells us that this is the number of columns of N that we need in order to generate, or compute, all the columns. We can think of such a smallest set of columns as one that captures within it the entire diversity amongst all the columns of N.

So which partitions $\{A, B\}$ have low order under this order function, and which have large order? For example, if some $b \in B$ are deemed similar to every $a \in A$ and the other $b \in B$ are similar to none, then the columns of N either consist of 1s only or of 0s only. Then the dimension of the space spanned by the columns of N is 1, since one well-chosen column suffices to compute all the others. (Indeed, let that column be one that consists of 1s; then the other columns are either identical to it or obtained from it by multiplying it with 0.)

This simple example shows that $|s|$ can have low order even if many $a \in A$ are similar to many $b \in B$, but the pattern of which a are similar to which b is a simple pattern. The function $s \mapsto |s|$ is submodular in the realm of all partitions of V [30].

Next, let us consider the $n \times m$ matrix from Section 9.3 that directly encodes the interaction between V and Q. This matrix M has $n = |V|$ rows indexed by the elements of V and $m = |Q|$ columns indexed by the elements of Q, with entries $v(q) = q(v) \in \mathbb{F}_2 = \{0, 1\}$ in row v and column q. Each row v displays the values of the function $v \colon Q \to \{0, 1\}$ that records all the answers given by v to the questions in Q, and each column q displays the values of the function $q \colon V \to \{0, 1\}$ that records all the answers to q given by the $v \in V$.

The rank of M, as before, is the number of columns we need in order to compute all the other columns. In the interpretation, this is low if from the answers to a few questions we can compute the answers to all the others in a simple way. But as earlier with N, the rank of M is equal also to the number of rows we need to compute all the other rows. In the interpretation, this is low if from the complete answers to Q of a few people in V we can compute the complete answers given by all the others, once more in the specific simple way of computation in $\mathbb{F}_2$.

Both interpretations of the rank of M can be rephrased informally as a degree of diversity of opinion regarding Q amongst the people in V. It is low if there exists a small group of 'influencers' in V whose views the other people in V follow. It is also low if there are a few 'key questions' in Q that sum up all the other questions between them.

Suppose now that we have two 0/1-matrices, M' and M'', both of whose columns are indexed by the elements of Q. Let us define as their *joint rank*, denoted as $\operatorname{rank}(M', M'')$, the smallest size of a subset P of Q such that two things happen at once: that from the columns of M'

indexed by the elements of P we can compute all the other columns of M', and simultaneously that from the columns of M'' indexed by the elements of P we can compute all the other columns of M''.

To define our next order function on the partitions of V, consider now a given such partition $\{A, B\}$. This divides M into two submatrices: the matrix M_A consisting of the rows of M whose index lies in A, and M_B consisting the rows whose index lies in B. The following three possible definitions of an order for $\{A, B\}$ suggest themselves:

$$|A, B| := \max (\operatorname{rank} M_A, \operatorname{rank} M_B)$$
$$|A, B| := \operatorname{rank} M_A + \operatorname{rank} M_B$$
$$|A, B| := \operatorname{rank} (M_A, M_B)$$

The second and third of these are submodular in the realm of all partitions of V, the first is not. Let us consider them in turn.

A partition $\{A, B\}$ of V has low order in terms of the first definition, where this is the maximum of the ranks of the submatrices M_A and M_B, if both these ranks are small: if the answers to a few well-chosen questions given by the people in A determine their answers to all the questions in Q, and likewise for B. Which these few questions are can be different for A and B, and here lies the main idea behind this order function: it considers $\{A, B\}$ as a natural split of V if the entire population of V maybe cannot agree on which are the essential questions in Q, but both the people in A and the people in B can agree on this amongst themselves, albeit with different results.

The second of the above order functions, which defines $|A, B|$ as the sum of the ranks of M_A and M_B rather than their maximum, focusses more directly on the agreement between the people in A and those in B. Compared with the (constant) overall rank of M it is large if their views on Q have a lot in common: if a lot of the diversity of opinion that can be found in V overall is present independently amongst the people in A and amongst the people in B alone. By contrast this $|A, B|$ is small, for example, if the people in A and in B are interested in different topics.

To make this more precise, let us pause our interpretation for a moment and take a closer look at the sum in the definition of $|A, B|$, comparing it with rank M. The linear algebra fact mentioned before tells us that the rank of a matrix not only equals the minimum number of columns needed to compute the other columns, as it does by definition, but also the minimum number of rows from which all the other rows can be computed. We can therefore compute all the rows of M whose index lies in A from rank M_A of these rows, and all the rows of M whose index lies in B from rank M_B of those rows. So we can compute all the rows of M from just $|A, B| = \operatorname{rank} M_A + \operatorname{rank} M_B$ rows. This shows that

$$\operatorname{rank} M \leqslant |A, B| \leqslant 2 \operatorname{rank} M$$

for this definition of $|A, B|$.

There are different readings of what is happening when $|A, B|$ is large, close to $2 \operatorname{rank} M$. One is that the diversity of opinion on Q is not much smaller among the people in A or B alone than in V overall: that we need nearly as many 'influencers' in A and in B to reflect the opinions held there alone as the number of influencers in $V = A \cup B$ that suffices to reflect the overall opinion on Q in V. Indeed, if this is the case then each of rank M_A and rank M_B is close to the rank of M, so their sum is closer to its upper bound of $2 \operatorname{rank} M$ than to its lower bound of rank M. Or, expressed with reference to dimension: $|A, B|$ is large if the spaces of 'diversity of opinion' spanned by A and by B do not fit nearly disjointly into the overall space for V, but have a large overlap.

Viewed dually, $|A, B|$ is large if we need nearly as many key questions to summarize the views of the people in A or in B alone as we do to summarize the views of all the people in V: both A alone and B alone reflect nearly the full spectrum of opinion that can be found in V.

Another aspect of $|A, B|$ being large is that the rank of M is a lot smaller than rank $M_A + \operatorname{rank} M_B$. This in turn has two aspects. One is that it helps to use rows with an index in B when computing the rows with an index in A, i.e., that in order to represent the views of the people in A it can help to ask people in B, and vice versa. The second aspect is that no matter how we pick a smallest set of influencers in all of V for the opinions held there, its subsets in A and B, respectively, will not both fully represent the diversity of opinion held there.

Conversely, suppose the people in A and in B are interested in different topics reflected by the questions in Q. Let us show that $|A, B|$ is small in this case, close to the rank of M. Indeed, if we need n_A influencers in A to reflect the diversity of opinion on Q prevalent in A, and n_B influencers for B, we will need nearly $n_A + n_B$ influencers in V to reflect the overall diversity on Q: this is because people in A cannot help us to represent views held in B, and vice versa, since their interests differ. So no matter how we choose our set of influencers for V, we will need at least about n_A people in this set only to represent the views held in A, because these people will have to belong to A, and n_B people different from these (because they will have to be in B) only to represent the views held in B.

Finally, there is our third definition of an order function: that $|A, B|$ is the joint rank of M_A and M_B, the smallest number of columns of M whose restrictions to A can generate all the columns of M_A and whose restrictions to B can simultaneously generate all the columns of M_B. In the interpretation, this is the smallest set of questions in Q that can serve as key questions both for the opinions held in A and for the opinions held in B.

At first glance, this looks like a description of the rank of M itself, the least number of columns of M from which we can compute all the other columns. And indeed, if we have such a set $P \subseteq Q$ of columns of M, then from their restrictions to A we can compute the other columns' restrictions to A, and similarly for B. But this only shows that our P from the definition of rank M is a *candidate* for the P sought in the definition of the joint rank. So all we have shown is that

$$\operatorname{rank}(M_A, M_B) \leqslant \operatorname{rank} M.$$

Can this inequality be strict? In other words, can there be a set $P \subseteq Q$ of questions that generates the full diversity of opinion on both A and B but not on $V = A \cup B$?

Surprisingly, there can. The reason lies in the fact that the notion of 'generate' is more flexible when applied separately to A and B, as it is in the definition of the joint rank, than when it is applied globally to V.

For example, let $P \subseteq Q$ be a smallest set of key questions for all of V, and pick one question p from this set. Then, by the minimality of P, we cannot compute the answers to p given by the people in V from their answers to the other questions in P: if we could, then p would not be needed as a key question. But it may happen that the views on p of the people in A can be computed from their views on the rest of P, while at the same time the views on p of the people in B can be computed from their views on the rest of P, since the ways in which these computations are performed may now differ.[15] In this case we can omit p from P for the definition of rank (M_A, M_B), but not for the definition of rank M. Thus, rank $(M_A, M_B) < \operatorname{rank} M$ in this example.

So when is rank (M_A, M_B) large? It will certainly be large as soon as one of M_A and M_B has large rank, as clearly

$$\max\left(\operatorname{rank} M_A, \operatorname{rank} M_B\right) \leqslant \operatorname{rank}\left(M_A, M_B\right).$$

But the converse is not true: it can happen that both M_A and M_B have small rank, because each has their own dedicated small set of key questions, P_A and P_B, say, but their joint rank is large since we cannot find a set $P \subseteq Q$ that works for both A and B at once, as $P = P_A = P_B$. Thus,

$$\begin{aligned}
\tfrac{1}{2}\operatorname{rank}\left(M_A, M_B\right) &\leqslant \tfrac{1}{2}\operatorname{rank} M \\
&\leqslant \tfrac{1}{2}\left(\operatorname{rank} M_A + \operatorname{rank} M_B\right) \\
&\leqslant \max\left(\operatorname{rank} M_A, \operatorname{rank} M_B\right). \\
&\leqslant \operatorname{rank}\left(M_A, M_B\right).
\end{aligned}$$

Comparing these rank functions numerically is of little value, of course, since we can always add or subtract a constant to or from any of them. However the inequalities listed above can remind us of how the corresponding notions of bottleneck compare, which is the more relevant question. Since the three functions are so closely tied together, let us just keep one of them for future reference, and let us deduct rank M, a constant, to make it more intuitive, as discussed:

$$|A, B| := \operatorname{rank} M_A + \operatorname{rank} M_B - \operatorname{rank} M. \qquad (\text{O8})$$

As observed earlier, these values of $|A, B|$ lie between 0 and rank M, and this rank function is submodular in the realm of all partitions of V.[16]

In some settings, for example if the range of answers to the questions in Q is $\mathbb{R}$ rather than $\{0, 1\}$ and V is smaller than Q,[17] it can happen that many of the matrices considered above have full rank, in which case (O8) yields $|A, B| = 0$. In such a context we can measure the diversity of opinion in a set A of people in a different way: by the 'volume' of what the rows of M indexed by A enclose in the space $\mathbb{R}^m$ of all $Q \to \mathbb{R}$ functions.[18] Let us call this number the *volume of* A; formally, it is the $|A|$-dimensional volume of the parallelotope formed in $\mathbb{R}^m$ by the vectors in A.

The idea is that if the diversity of opinion amongst the people in A is small, i.e., if their answers given to Q are similar, then geometrically the lines from the origin of $\mathbb{R}^m$ to the points $v(Q) \in \mathbb{R}^m$ with $v \in A$ that encode these answers form small angles with each other, which makes the volume of A small. If this happens for both A and B in a partition $\{A, B\}$ of V, then this is a good partition, which should be assigned low order – especially if the diversity of opinion *between* A and B is larger. So the idea is to define $|A, B|$ as something like the sum, or better[19] the product, of the volumes of A and B.

More formally, recall the matrix $-J := MM^\top$ from Section 9.3. This is an $n \times n$ matrix indexed by the elements of V. For any $A \subseteq V$ denote by J_A the submatrix of $-J$ that consists of the rows and columns indexed by elements of A. Note that $J_A = M_A M_A^\top$.

Linear algebra tells us that the volume of A is the square root of the determinant $\det J_A$.[20] Our idea to define $|A, B|$ in terms of the volumes of A and B thus led us to consider the possible definition of

$$|A, B| := (\det J_A)(\det J_B).$$

This is small if the volumes of A and B are small, as intended.

Interestingly, the number $\det J_A$ is also the volume (not its square) of the parallelotope spanned in $\mathbb{R}^{|A|}$ by the rows of J_A; let us call this the *J-volume of* A. Recall that the rows of J_A are the $|A|$-vectors that list, for the row indexed by $u \in A$ say, the correlations $\langle u, v \rangle$ of u with the other $v \in A$ in $\mathbb{R}^m$, i.e., as $Q \to \mathbb{R}$ functions. Two such rows, indexed by u and u' say, form a small angle in $\mathbb{R}^{|A|}$ if $\langle u, v \rangle$ is close to $\langle u', v \rangle$ for many $v \in A$: if the opinions of u and u' differ from the views of the other $v \in A$ to similar extents (though not necessarily in similar ways). This happens, of course, whenever u and u' themselves hold similar views. The converse implication is less clear, which contrasts somewhat with the fact that the J-volume of A is so closely related to its ordinary volume, being equal to $\det J_A$ and thus to the square of the volume of A.

The order function just discussed is not submodular. However, we can make it so by taking its logarithm. Indeed, the function $\log \det J_A$ on the subsets of V is known to be submodular [27], and hence so is the order function

$$|A, B| := \log \det J_A + \log \det J_B - \log \det J, \tag{O9}$$

where the constant of $\log \det J$ is deducted for normalization purposes, just as we deducted rank M in (O8). However it also has other beneficial effects, as we shall see in a moment.

Note that these logarithms are defined only if the determinants are non-zero. To ensure this, we apply (O9) only if V is linearly independent as a subset of $\mathbb{R}^m$.[21] Then its subsets A and B are linearly independent too, all determinants above are non-zero, and the logarithms are defined.

Crucially for us, V cannot be linearly independent if $|V| > |Q|$, which in a classical questionnaire scenario is the rule rather than the exception. Although the case that V is linearly independent (so that all the matrices M_A and M_B have full rank) motivated us to consider (O9) as a more subtle alternative to our earlier rank-based order function of (O8), the assumption that V is linearly independent, and in particular that $|V| \leqslant |Q|$, remains a serious caveat for now; we shall return to it later.

Compared with the values of rank M_A considered earlier, the values of $\det J_A$, and even of $\log \det J_A$, can depend rather drastically on single elements of A. Indeed, when we add a single new element $a \notin V$ to A, the volume of A changes by a factor $h(a, A)$ that can be arbitrarily close to zero: it is the *height* of a over the subspace of $\mathbb{R}^m$ generated by A. In the extreme case (which our requirement that V be linearly independent rules out) that a already lies in this subspace, this even reduces the volume of A to zero. This can happen no matter how large the volume of A was before we added a: depending only on where a lies relative to A, it can be reduced to almost zero.[22]

Expressed in terms of $\det J_A$ rather than volume, adding a to A multiplies $\det J_A$ by a factor of $h(a, A)^2$ (which can be close to zero), and adds $2 \log h(a, A)$ (which can be very negative) to $\log \det J_A$. In (O9), however, this is offset by a similar effect on $\log \det J$, so that the overall value of $|A, B|$ is always non-negative [25]. Here lies the main benefit of subtracting the constant of $\log \det J$ in (O9).

Another effect lies in how it affects our earlier comparison between related partitions of V, when we changed V a little by adding a new element $a \notin V$ to A. One can prove that this will no longer reduce $|A, B|$. It leaves $|A, B|$ unchanged if the point closest to a in the span of V in $\mathbb{R}^m$ is spanned by A alone, and will increase it the more the greater the influence of B is in generating that point.[23]

The order function in (O9) thus encourages us, when we try to build a bottleneck partition $\{A, B\}$ of V by adding one point of V to either A or B at a time, to add any new point for V to the side, A or B, whose elements it resembles more closely in terms of how they think about Q.

Let us return now to the question of how to adapt (O9) when V is not linearly independent, and hence $\det J_A = 0$ for some $A \subseteq V$. One possibility is to replace $\det J_A$ in (O9) with the positive numbers $\det J_{A'}$ for suitable linearly independent subsets A' of A (and likewise for B). Alternatively, we might try to replace $\det J_A$ with the product of just the non-zero eigenvalues of J_A,[24] or replace $\log \det J_A$ with the sum of their logarithms.

Such more sophisticated order functions are explored in [25]. The ideas just mentioned should be understood not so much as suggesting off-the-shelf order functions to suit all purposes, but rather as ingredients for the design of order functions tailored to the intended tangle application.

A similar approach can be taken with other positive semi-definite matrices whose rows and columns are indexed by V and which in position (u, v) express a degree of similarity between the elements $u, v \in V$. The Laplacian L, for example, decomposes as $L = WW^\top$ as defined in Section 9.2. The *unsigned Laplacian* L_+, obtained from L by replacing its off-diagonal entries $-\sigma(u, v)$ with $\sigma(u, v)$, decomposes as $L_+ = W_+ W_+^\top$, where W_+ is obtained from W by replacing all its entries by their absolute values.[25]

Writing L_A for the submatrix, of either L or L_+, whose rows and columns are indexed by the elements of A, we can consider the following analogue of (O9):

$$|A, B| := \log \det L_A + \log \det L_B \qquad (\text{O9}')$$

whenever these determinants are non-zero. This is the case, for example, as soon as $\sigma(u, v) \neq 0$ for all distinct $u, v \in V$; see Section 9.2.[26] We cannot subtract $\log \det L$ here in analogy to (O9), since the determinant of L is always zero. The determinant of L_+ can be non-zero, in which case we may wish to subtract $\log \det L_+$ in (O9'). The order function in (O9') is submodular for either L or L_+ [27].

9.5 Entropy-based order functions

In this section we introduce order functions on set partitions which combine the two aspects that motivated us in the last section: the complexity of the internal structure on each side of the partition, and that of the structure across it, of how the two sides interact. While these were two separate aspects in the last section, we now combine them into one: the order functions we shall be looking at measure how the internal structures of the two sides of a partition compare with each other. Very roughly, the idea is that if these structures have nothing to do with each other, we think of this partition as good and assign it low order; if they are similar, then this partition is a bad way to cut our set and will receive high order.

There are two ways in which this section can be read. Thanks to the very intuitive terminology used in the mathematical theory of discrete entropy, it is possible to ignore all the formal mathematics and still get some idea of what the order functions defined here measure. On the other hand, for any reader who wishes to understand them at a mathematical level, the intuitive terminology can hinder as much as it can help: it presents a temptation to forget the formal definitions too quickly, replacing them in one's thinking with the intuition that we have for the words chosen for these definitions (such as 'mutual information'). Like the rest of this book, this section can be read with a minimum of formal mathematical training but enough care to detail. Such careful reading will be slow, but given enough time it should be easy.

Before we can define our order functions, we need some preparation. Once more, let us assume that our data on V is given as in the standard questionnaire setting: in the form of a set Q of functions on V, with values in $\{\text{yes}, \text{no}\}$ or in $\{-5, \ldots, 5\}$, which we think of as questions answered by all the $v \in V$.

Let $f\colon V \to X$ be any function from our set V to some set X. For a start, we may think of f as a question q in Q and of X as the set of answers given to the questions in Q, i.e. $X = \{\text{yes}, \text{no}\}$ or $X = \{-5, \ldots, 5\}$. Later, these f can also correspond to pairs of questions, or to partitions $\vec{s}$ of V to which we wish to assign an order, in which case we take $X = \{0, 1\}$ or $X = \{-1, 1\}$ to encode such partitions as functions on V.

Given f and some fixed $x \in X$, let us define our *surprise* at seeing f take the value x, or, in our example, at seeing f answered as x, as the number $1/p_f(x)$, where

$$p_f(x) := |\{\, v \in V \colon f(v) = x \,\}|/|V|$$

is the proportion in V of those elements v that answered f as x. In probabilistic language,[27] it is the probability that a randomly chosen element of V answers f as x. Thus, if few v in V are such that $f(v) = x$, our surprise at seeing this answer for such a 'random' v will be great, while if most $v \in V$ answer f as x it will be small.

Suppose, for example, that $X = \{\text{yes}, \text{no}\}$, and that half the people in V answered a question $q \in Q$ as 'yes', the other half as 'no'. Then $p_q(\text{yes})$ and $p_q(\text{no})$ are both $1/2$, and our surprise at seeing either one of them will be 2. If all the people in V answered q as 'yes', then $p_q(\text{yes}) = 1$, and our surprise at seeing q answered as 'yes' will be $1/1 = 1$. Our surprise at seeing it answered as 'no' will be formally $1/(0/|V|) = 1/0$, and thus be undefined. This is a good thing: it would make no sense to have a mathematical expression for how surprised we would be at seeing something impossible, and the fact that $1/0$ is undefined in mathematics nicely reflects this. However, we can quantify our surprise at seeing q answered as 'no' if very few people in V chose that answer: it will tend to infinity as the proportion of $v \in V$ for which $q(v) = \text{no}$ tends to zero.

Suppose now that we have two questions $q', q'' \in Q$, two answers $x', x'' \in X$, and we know both our surprise at seeing q' answered as x' and our surprise at seeing q'' answered as x''. What is our surprise at seeing both happen, i.e., of seeing q' answered as x' and q'' answered as x''?[28] Well, it depends. If $q' = q''$ and $x' = x''$, for example, it will be no greater than our surprise at seeing one of them happen, since the other will then also happen. At the other extreme, if $q' = q''$ and $x' \neq x''$, it will be impossible that both $q'(v) = x'$ and $q''(v) = x''$ for any v, and our surprise at seeing both happen will be undefined.

Between the two extremes, suppose that q'' divides each of the two sides of the partition of V given by the answers to q' with the same proportions as it divides the entire V, and similarly with the roles of q' and q'' swapped. We then say that q' and q'' are *independent* of each other. Figure 9.1 shows an example of independent q', q'' with different proportions. For such independent q' and q'', the surprise at seeing both q' answered as x' and q'' answered as x'' will be the product of the two individual surprises.[29]

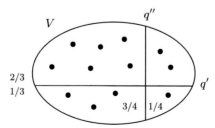

FIGURE 9.1. Each of the questions q', q'' divides the other's answer sets with the same proportions as it divides V.

This does not quite match our intuitive use of the word 'surprise', which is more additive than multiplicative.[30] In order to better align the definition of 'surprise' with our intuitive use of the word, let us define the *logarithmic surprise* at seeing a function $f\colon V \to X$ take the value x as the binary logarithm of its ordinary surprise, as $\log\bigl(1/p_f(x)\bigr)$. The logarithmic surprise at seeing independent functions both take some given values, then, is the sum of the surprises at seeing each of them take that particular value, rather than their product.

Our earlier examples for $X = \{\text{yes}, \text{no}\}$ adapt to this change as follows. If half the people in V answer a question q as 'yes' and the other half as 'no', then the logarithmic surprise of each of these is 1, the logarithm of their ordinary surprise (which was 2). If all the $v \in V$ answered q as 'yes', the logarithmic surprise at seeing this will now be 0, the logarithm of its ordinary surprise of 1. This is in line with our intuitive use of the word 'surprise': if we know that everybody in V answered q as 'yes', there will be no surprise in learning that a 'random' $v \in V$ answered q as 'yes'. The logarithmic surprise at seeing q answered as 'no' will still be undefined, which also matches our intuition. And if the proportion of $v \in V$ that answer q as 'no' is positive but we let it tend to zero, the logarithmic surprise at seeing this will still tend to infinity, albeit more slowly than the ordinary surprise does.

Finally, let us consider for any fixed question q the average, taken over all the possible answers $x \in X$, of our (logarithmic) surprises at seeing q answered as x. Not the absolute average, but the average weighted for each x with the proportion of people that chose x as their answer to q. More precisely, we define as the *expected surprise* for a function $f\colon V \to X$ the number

$$H(f) = \sum_{x \in X} p_f(x) \cdot \log\bigl(1/p_f(x)\bigr).$$

This is a non-negative real number known as the *discrete entropy* of f. Its smallest possible value is 0, and it can get arbitrarily large if X does, too. *Note that $H(f)$ does not depend on the actual values of the function f, only on the relative[31] sizes of the sets $f^{-1}(x)$ of elements of V it sends to the various values $x \in X$.*

Let us go back to our earlier examples with $X = \{\text{yes}, \text{no}\}$. If half the people in V answer a question q as 'yes' and the other half as 'no', then the logarithmic surprise of each of these was 1, so their weighted average $H(q)$ is also 1.

More interestingly, how does the expected surprise for q behave when the proportion of $v \in V$ that gave one answer, say 'yes', tends to 1 and the other to 0? As we saw earlier, the logarithmic surprise at seeing q answered as 'yes' will tend to 0 then. But the logarithmic surprise at seeing q answered as 'no' will tend to infinity. So what is the limit of their weighted average, the expected surprise $H(q)$ for q?

It turns out that this is zero too. Indeed, the decline of the weight factor $p_f(x)$ under the sum in the definition of $H(f)$ for $f = q$ and $x =$ 'no' outweighs the growth of the logarithmic factor $\log(1/p_f(x))$ next to it as $p_f(x)$ tends to zero (and $1/p_f(x)$ tends to infinity), so their product tends to 0. In words: although our surprise at seeing q answered as 'no' grows as this gets less and less likely, the increase of our surprise is not big enough to offset the decrease of the chance of this happening.

Thus, both summands in $H(q)$ tend to 0, and hence so does $H(q)$.

Let us return to our earlier example where we considered two yes/no questions $q', q'' \in Q$. Now $H(q'')$ formalizes our expected surprise at learning the answer to q'' of some unkown $v \in V$. Does this change if we are told how v answered q'? Well, it depends on how q' and q'' are related. If $q'' = q'$, for example, then knowing anyone's answer to q' tells us their answer to q'', leaving us with no surprises for q''. Interestingly, the same is true if q'' is very *unlike* q': if it is its negation.

Between these two extremes, if q' and q'' are independent, then knowing the answer to q' should have no influence on our surprises for q''. To make this more precise, let us say that 'knowing the answer' to q' means that we restrict our attention when considering q'' to only those $v \in V$ that gave this particular answer to q'. If q' and q'' are independent, then such a restriction should indeed have no effect on our surprises for q'', since q'' divides each side of the partition q' of V as it divides V.

More generally, consider two functions $f_1: V \to X$ and $f_2: V \to Y$. Let us define as the *expected surprise for f_2 given the value of f_1* the number

$$H(f_2|f_1) := \sum_{x \in X} p_{f_1}(x) \cdot H\left(f_2 \restriction f_1^{-1}(x)\right)$$
$$= \sum_{x \in X} p_{f_1}(x) \cdot \sum_{y \in Y} p_{f_2 \restriction f_1^{-1}(x)}(y) \cdot \log\left(1/p_{f_2 \restriction f_1^{-1}(x)}(y)\right),$$

where $f_2 \restriction f_1^{-1}(x)$ is the restriction of f_2 to $f_1^{-1}(x) \subseteq V$, the set of those $v \in V$ which f_1 sends to x.

Formally, this is known as the *conditional entropy* of f_2 given f_1. Despite its look, the formula is quite simple once we have deciphered its notation, and it expresses exactly what we intended. The second sum, complicated though it looks, is simply the expected surprise for the function f_2 restricted to the subset $f_1^{-1}(x)$ of V: the expected surprise for f_2 when we consider only those $v \in V$ that f_1 sends to x. Of course, V is the disjoint union of these subsets $f_1^{-1}(x)$, one for every $x \in X$ that occurs in the image of f_1. The entire expression $H(f_2|f_1)$, therefore, is simply the weighted average over all $x \in X$ of those expected surprises for f_2 restricted to $f_1^{-1}(x)$, the weight for each x being the probability that this x occurs as the value of f_1 (Figure 9.2).

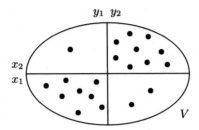

FIGURE 9.2.   Restricted to either of the two values x_1, x_2 of f_1, the entropy of f_2 is lower than it is on all of V.

If we re-calculate our earlier examples for two questions $q', q'' \in Q$ formally, we find that the conditional surprises we just defined behave exactly as envisaged. For example, if $q'' = q'$ or q'' is the negation of q', then $H(q''|q') = 0$: either answer to q' tells us the answer to q'', leaving us with no surprises there. If q' and q'' are independent, on the other hand, then $H(q''|q') = H(q'')$: knowing the answer to q' does not help us guess the answer to q''.[32]

More generally one can show that $H(q''|q') \leqslant H(q'')$ for all q' and q'', indeed that $H(f_2|f_1) \leqslant H(f_2)$ for all functions f_1, f_2 on V. This reflects our intuition that knowledge about f_1, no matter how irrelevant it may seem, can never harm our guesses for f_2, only help. The difference between the two quantities, the number $H(f_2) - H(f_2|f_1)$, measures how much it helps: how much f_1 tells us about f_2.[33]

It turns out that this difference is symmetric in f_1 and f_2: knowing f_1 helps us with guessing the values of f_2 exactly as much as the other way round. We call this common value the *mutual information*

$$I(f_1, f_2) := H(f_1) - H(f_1|f_2) = H(f_2) - H(f_2|f_1)$$

of f_1 and f_2, and think of it as how much knowing a value of one of these two f_i tells us, on average, about the other's values – or more precisely, about how the subsets of V on which the other f_i takes its values are distributed.

For example, consider again our two yes/no questions q', q'' answered by the people in V. If $q' = q''$ or one is the negation of the other, then $H(q'|q'') = 0 = H(q''|q')$, so $I(q', q'')$ is at its maximum: the answers to either of q' or q'' tell us everything there is to learn about the other. At the other extreme, if q' and q'' independent, then $I(q', q'') = 0$: we cannot learn anything from the answers given to either of them about the distribution of answers given to the other.

We are now ready to define some entropy-based order functions. The simplest of these is not yet an order function defined on all the partitions s of V, as envisaged, but one on $Q: V \to X$ itself: let

$$|q| := H(q) \tag{O10}$$

for all $q \in Q$. This measures how the various answers to q divide the people in V: it will be low if just a few of the possible answers attracted most of the people in V, and it will be large if V is evenly divided into the sets $q^{-1}(x)$ for $x \in X$.

Since $H(f)$ is defined for all $V \to X$ functions f, our definition of $|q|$ extends readily to combined questions, such as suprema $q = q' \vee q''$ and infima $q = q' \wedge q''$ of questions $q', q'' \in Q$ when $X = \{-5, \ldots, 5\}$ (say), as defined in Section 7.6. Since H is not submodular on $\{-5, \ldots, 5\}^V$ as a function, we are not guaranteed a tree-of-tangles theorem on Q and the functions in $\{-5, \ldots, 5\}^V$ obtained from Q, by taking infima and suprema, of order less than any fixed k we may specify. But adding such suprema and infima locally as needed can make Q submodular as a feature system, and we may still get a tree-of-tangles theorem in practice.

Next, let us see some entropy-based order functions defined on all the partitions s of V. To get started, we pick for every such s a default orientation $\vec{s}: V \to \{-1, 1\}$ that assigns 1 to the elements of one of the two sides of s and -1 to those of the other side.

Similarly, we think of Q as the function that sends each $v \in V$ to the family $(v(q))_{q \in Q}$ of answers it gave to the questions in Q.[34] If the set of possible answers is $\{-5, \ldots, 5\}$, then Q defines a map $V \to \{-5, \ldots, 5\}^Q$.

We now define

$$|s| := H(Q|\vec{s}) \tag{O11}$$

as the *global entropy order* of s with respect to Q. This is well defined, since $H(Q|\vec{s}) = H(Q|\overleftarrow{s})$ by the definition of conditional entropy. Note also that Q and $\vec{s}$ are functions with the same domain, V, as required in the definition of H. If desired, we can rewrite $|s|$ in terms of mutual information:

$$|s| = H(Q) - I(Q, \vec{s}). \tag{O11$'$}$$

Either way, the interpretation of $|s|$ is that it measures the gain in structure when we shift our attention from how Q divides V to how it divides the sides of s, its bigger side being more important. For example, if Q contains questions about two different topics, and half the people in V are interested in one of these while the rest are interested in the other, then letting s divide V into these two sets of people can give it low order if the questions are phrased suitably.[35]

Indeed, when people are interested in a topic, their views on questions about it are likely to be more pronounced and less random than when they have no knowledge or tastes on this topic. If the questionnaire is designed in such a way that a structured variety of a few pronounced views, as opposed to a chance distribution of random views, translates to a similarly structured partition of V defined by the various answers given to Q, then the entropy of Q will be smaller on sets of people interested in its topic.

Hence if s divides V roughly according to interest in one of two topics, then the order of s, the (weighted) average of the entropies of Q restricted to its two sides, will be lower than that of Q applied to all of V.

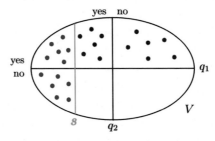

FIGURE 9.3.   The partition s cuts V into equally sized sets of red and blue elements. $H(Q) = 3/2$, but $H(Q|\vec{s}) = 1$.

Figure 9.3 shows the simplest case of this example, where s divides V into sides of equal size, one red and one blue, while Q consists of two questions, q_1 and q_2. The people on the red side $\vec{s}$ of s agree on question q_1 but are indifferent to question q_2; the people on the blue side $\overleftarrow{s}$

agree on question q_2 but are indifferent to question q_1. Overall, the entropy of Q decreases from $3/2$ to 1 when we restrict its domain from V to either side of s, so its average $H(Q|\vec{s})$ of these two is also 1.

If our data comes from a big questionnaire Q answered by a small population V, it can happen that the sets $Q^{-1}(z)$ of people that gave any particular set $z \in \{-5, \ldots, 5\}^Q$ of answers can be quite small, even singletons. This can happen, for example, if Q contains many questions that most of the people in V are not interested in, which are thus likely to split V about equally in a random way. If this happens, s cannot partition these sets further. Then $|s|$ will have nothing to do with our data as encoded in Q: we simply have $|s| = H(\vec{s})$. In particular, the order $|s|$ of s will not tell us whether s is a good partition of V to base our tangles on, which was the whole point of considering order functions.

In such cases, it may be a good idea to evaluate the effect of splitting V by s not on the entire set Q of questions but on individual questions $q \in Q$. Let us define the *local entropy order* of s with respect to Q as

$$|s| := \sum_{q \in Q} H(q|\vec{s}). \tag{O12}$$

This order function will reward s, by making $|s|$ smaller, if restricting many questions q from V to one of the sides of s decreases their entropy a lot (because many people *on that side* agree in their answers to q). This effect will still be visible when Q also contains many questions answered randomly by many people in V, which would kill the usefulness of our global entropy order function.

Note the similarity of the local entropy order to the order function (O7) we defined in Section 9.3. There, we assigned low order $|f|$ to an advanced feature $f: V \to \mathbb{R}$ if f was highly correlated to many $q \in Q$. The difference now is that s need not have the same format as the elements of Q: it can be a map $V \to \{\text{left}, \text{right}\}$ while the questions in Q take values in $\{-5, \ldots, 5\}$. But the two remain similar in their approach. Their difference comes down to the difference between the correlations $\langle q, \vec{s} \rangle^2$ and $H(q|\vec{s})$ of the questions q with $\vec{s}$. Note that both are invariant under replacing either q or $\vec{s}$ with their negative, the negation $-q$ of q or the inverse $\overleftarrow{s}$ of $\vec{s}$.

Neither the global nor the local entropy order function is submodular.

9.6 Entropy-based order functions for the dual system

In this section we look at another way to define order functions on the partitions of a set in terms of entropy. As before, we compare the entropy of the data encoded on one side of the partition with what is encoded on the other side, and define the order of the partition as the mutual information between the two. Thus, a partition will receive low order, encouraging its adoption for our tangles, if we can learn little from how the data structures one side about how it structures the other side.

Our plan now is to do this with partitions of the set Q of questions rather than of the set V of people. From a mathematical point of view, this is immaterial: as pointed out in Section 7.7, any given feature system can be viewed as primal or as the dual of another system (namely, of its dual), so the two forms can be translated into each other. However, the order functions we shall consider in this section are not merely translations of those we studied in Section 9.5 by dualization.

Moreover, when we partition questions rather than people, entropy terms such as 'mutual information' are more intuitive. We shall also apply entropy-based order functions for partitions of Q in Section 9.7, so it will be convenient to introduce them already in this form.

As in Section 9.5, we think of a subset Q' of Q as the function that maps each $v \in V$ to the collection x of answers it gave to the questions in Q'. If the questions in Q took answers in $\{-5,\ldots,5\}$, say, then the set of possible answers is $X = \{-5,\ldots,5\}^{Q'}$, the set of all maps from Q' to $\{-5,\ldots,5\}$. So in this case $Q'\colon V \to X$ is given by

$$Q'\colon v \mapsto (v(q))_{q\in Q'} \in \{-5,\ldots,5\}^{Q'}.$$

We now define the *global entropy order* of a partition $\{Q_1,Q_2\}$ of Q as the mutual information of the functions Q_1 and Q_2, as defined in Section 9.5:

$$|Q_1,Q_2| := I(Q_1,Q_2). \tag{O13}$$

This is well defined, since $I(Q_1,Q_2) = I(Q_2,Q_1)$.

How can we picture this mutual information of Q_1 and Q_2? Much the same as $I(q_1,q_2)$ for single questions q_1 and q_2. The only difference now is that the functions Q_1 and Q_2 partition V into more than two sets: into the answer sets $Q_1^{-1}(x)$ for all the various $x \in \{-5,\ldots,5\}^{Q_1}$, and likewise for Q_2. Broadly speaking, $|Q_1,Q_2|$ tends to be high if these sets are similar for Q_1 and Q_2: if they partition V in similar ways. Note that this has nothing to do with what the questions in Q_1 and Q_2 mean: all that matters are the patterns into which they divide the population V.

When we look at the definition of $|Q_1, Q_2|$ more closely, we see that it focusses on how the two functions Q_1 and Q_2 partition each other's partition sets compared with how they partition V. For example, if Q_1 and Q_2 partition V into similar sets, they will *not* cut up each other's partition sets substantially, because every partition set for Q_1 will be similar to one for Q_2, and vice versa. In this case, the mutual information of Q_1 and Q_2 will be large.

More generally if Q_2, say, imposes more structure on a typical set $Q_1^{-1}(x)$ than it does on V, then the answers to Q_1 tell us a lot about the answers to Q_2, and $|Q_1, Q_2|$ will be large. By contrast, if Q_1 and Q_2 partition each other's partition sets with similar proportions as they each partition V, then $|Q_1, Q_2|$ is low: we cannot learn much about the answers to Q_2 from the answers to Q_1, or vice versa.

The entropy order can differ markedly from the similarity-based order functions on the partitions of Q that we considered earlier in this chapter (albeit with the dual notation), where $|Q_1, Q_2|$ was a sum of similarites $\sigma(q_1, q_2)$ taken over the pairs $q_1 \in Q_1$ and $q_2 \in Q_2$. Our earlier approach was to consider q_1 and q_2 as similar if many $v \in V$ gave them similar answers, so that splitting such q_1 and q_2 by our partition $\{Q_1, Q_2\}$ would increase its order, the more so the greater the similarity between q_1 and q_2.

The mutual information of Q_1 and Q_2, our new order $|Q_1, Q_2|$, is also likely to increase when many such pairs are split, because similar q_1, q_2 carry a lot of mutual information that drives up $I(q_1, q_2)$. However, $I(q_1, q_2)$ is also large when q_1 and q_2 are very dissimilar in terms of $\sum_{v \in V} \left(C - (q_1(v) - q_2(v))^2 \right)$, say, e.g. when $q_2 = -q_1$ is the negation of q_1. While splitting such pairs was not strongly discouraged by our earlier order functions, it is discouraged by the entropy order of $\{Q_1, Q_2\}$. This effect is even more immediate for the *local entropy order* of $\{Q_1, Q_2\}$ defined as

$$|Q_1, Q_2| := \sum_{q_1 \in Q_1} \sum_{q_2 \in Q_2} I(q_1, q_2). \tag{O14}$$

In applications where the wording of the questions in Q was not chosen carefully to distinguish between a question and its negation, but these were considered as essentially the same question, the entropy order functions, local or global, may thus be exactly what we want. Low-order tangles of the partitions of Q, in our questionnaire scenario, would typically represent something like the topics of the questions in Q, or groups of topics that show some correlation in how they divide the population of V. When such correlation is not due to any similarity in meaning, it will be an interesting outcome of a tangle analysis of V and Q.

As already discussed in Section 9.5, a technical advantage of the local over the global entropy order in some applications is that the global partition sets $Q_1^{-1}(x)$ for the various $x \in \{-5, \ldots, 5\}^{Q_1}$, say, can be quite small, even singletons. If this happens, Q_2 cannot partition these sets further. Then Q_2 will be constant on those sets so that $H(Q_2|Q_1) = 0$, resulting in $I(Q_1, Q_2)$ taking the maximum possible value for all such partitions $\{Q_1, Q_2\}$ of Q. In such cases, the local entropy order may be more meaningful than the global. It is also possible to mix the global and the local approach, by summing terms $I(Q_1', Q_2')$ for small subsets $Q_1' \subseteq Q_1$ and $Q_2' \subseteq Q_2$, or subsets consisting of questions deemed particularly important.

Both the global and the local entropy order function are submodular on the set of all partitions of Q.

9.7 Order functions based on tangles of the dual system

Recall that, in our simplest scenario of a system $\vec{S}$ of binary features of some dataset V, a tangle of S is a collection of 'typical' features, features that are often found together in the elements of V. When S took the form of a questionnaire Q, a tangle of S would be one particular way of answering all the questions, a way typical for some of the $v \in V$. We might thus think of a tangle of this feature system as something like a typical, if hypothetical, phantom element of V representing many others.

Likewise, a tangle of the dual of such a feature system, defined in Section 7.7, is one particular way of assigning to every $v \in V$ a 'yes' or a 'no' answer: a way that is typical for some of the questions in Q. We may thus think of the tangles of the dual system as typical, if hypothetical, phantom elements of Q representing many others.

Now the elements of Q, the questions in a questionnaire giving rise to a feature system $\vec{S}$, played a central role in how we computed the order of a partition $\{A, B\}$ of V in Section 9.1: we defined this as

$$|A, B| := \sum_{a \in A} \sum_{b \in B} \sigma(a, b), \tag{O1}$$

where $\sigma(a, b)$ counts the $q \in Q$ to which a and b give the same answer. In Section 9.2 we generalized this for advanced features $f: V \to \mathbb{R}$ to

$$\sigma(u, v) := \sum_{q \in Q} \left(C - (q(v) - q(u))^2 \right),$$

based likewise on questions $q: V \to \mathbb{R}$, where $q(v) = v(q)$ is v's answer to q and C is just a constant designed to ensure that the sum is non-negative. And even the order function for advanced features we discussed in Section 9.3 could be rewritten in terms of similarities between elements u, v of V expressed as a sum over the elements of Q, as

$$\langle u, v \rangle = \sum_{q \in Q} q(u) \cdot q(v).$$

The new idea now is to replace, in the definition of $\sigma(u, v)$ or $\langle u, v \rangle$, the index set Q in the sum with a suitable set of tangles, such as the k-tangles for some small k, of the dual feature system: typical, if virtual, 'questions'. Since all these tangles define functions $V \to \mathbb{R}$, just as the original $q \in Q$ do, the terms $q(u)$ and $q(v)$ in the definition of $\sigma(u, v)$ or $\langle u, v \rangle$ will be defined also when q is such a tangle.

However, with this new definition of similarities, a partition of V will be penalized for splitting a pair u, v only whenever they differ on such a tangle, a hypothetical question typical for Q, not on just any question in Q. In a standard interpretation, these dual tangles might indicate the main topics addressed in the questionnaire Q, or some structure on the set V of people that answered Q, such as their social background.

In a more refined approach one might take these tangles with weights that reflect just how typical they are for the $q \in Q$, or with weights that take account of their importance as reflected in the value of k.

The use of k-tangles requires, of course, that we already have an order function in place on the dual feature system. That order function should be defined in a different way, to avoid circularity in our definitions. In a questionnaire scenario, our entropy-based order functions from Section 9.6 fit this bill perfectly: they are not similarity-based, and when applied to partitions of Q rather than of V, the technical meaning of 'mutual information' between the sides of a partition coincides with its natural meaning particularly well.

This suggests that the various definitions of order functions from the first three sections of this chapter might yield even more interesting results when their various sums are taken not over the questions in Q themselves, but over the tangles of the dual feature system that represent these questions.

9.8 Tweaking order functions to increase balance of partitions

Some of the order functions we have discussed, including the most obvious ones, (O1) and (O5), tend to assign particularly low orders to unbalanced partitions of V, those that have only few elements of V on one side and most on the other. In the extreme case, which however is quite common in practice, the partitions of lowest order will pitch a single element of V against all the others. Can that be a problem?

At a theoretical level, it is not a problem. Most or all tangles will orient such unbalanced partitions towards their bigger side. This means that for low k we have only one tangle, and hence our k-tangles will get interesting only when k gets a little bigger. Put another way: since those unbalanced partitions do not distinguish any tangles, they have no bearing on the tangle structure of our data and will do neither good nor harm.

At a practical level, however, unbalanced partitions can clutter our feature systems and drain computing power, even and especially if their consideration has no material effect. This is a phenomenon well known in traditional clustering, and so a number of mechanical adjustments have been suggested to counteract it. We shall briefly mention a couple of these first, and then turn our attention to some of our more specific order functions.

Consider the order function (O1). It is based on a similarity function $\sigma\colon V^{(2)} \to \mathbb{R}$ and sums, given a partition $\{A, B\}$ of V, all the similarities $\sigma(a, b)$ with $a \in A$ and $b \in B$ (which we think of as 'penalties' to pay for cutting V here). If $\{A, B\}$ is balanced, this sum has about $|V|^2$ terms; if it is very unbalanced it has only linearly many, about $c\,|V|$ for some small positive constant c. And the sheer number of summands, even when these are small, tends to drive up the order of balanced partitions.

The simplest remedy for this bias against balanced partitions is not to take the sum of all these penalties $\sigma(a, b)$ but their average: to set $|A, B| := |s|/(|A| \cdot |B|)$, where $|s|$ is the order from (O1) or any other.

A similar way to generate more balanced partitions is to divide the given order $|s|$ separately by the cardinalities of A and B, and then to average over the two quotients: to replace $|s|$ with $\frac{1}{2}(|s|/|A| + |s|/|B|)$.

When $|s|$ is as in (O1), this tweak makes it constant with value $|V|/2$ if $\sigma(a, b) = 1$ for all a and b, and thus favours neither balanced nor unbalanced partitions. But, as before, $|s|$ could be any other order function. In graph clustering, this tweak is known as *ratio cut* [34].

Yet another way, which works for similarity-based order functions only, is to divide $|s|$ not by the cardinalities of A and B as above, but by $\sum_{a \in A} \sum_{a \neq v \in V} \sigma(a, v)$ and $\sum_{b \in B} \sum_{b \neq v \in V} \sigma(b, v)$, respectively. Each of these includes the value $\sum_{a \in A} \sum_{b \in B} \sigma(a, b)$ of the given $|s|$, but the tweak adds similarities inside the respective sides in the denominator to encourage partitions to keep similar elements together. This term is constant with value $|V|/(|V| - 1)$ when $\sigma(u, v) = 1$ for all $(u, v) \in V^2$, and thus again favours neither balanced nor unbalanced partitions. In graph clustering, this method is known as *normalized cut* [33].

None of these tweaked order functions is submodular, even if $|s|$ was. If keeping submodularity seems important, there is another simple way of dealing with the bias of (O1) towards unbalanced partitions: choose as σ a similarity function that decays fast, one that assigns *much* smaller values to dissimilar pairs u, v than to similar pairs. For example, if V is a set of points in the plane, we might take $\sigma(u, v) := 2^{c \cdot d(u,v)}$ where c is some negative constant and $d(u, v)$ is the Euclidean distance of u and v – rather than $1/d(u, v)$, say. In Figure 1.3, for example, this should identify the bottleneck consisting of the 'handle' as a good place for a low-order partition to cut: the quadratically many pairs (u, v) with u and v lying in different clusters left and right of the handle do not weigh in enough compared with the few pairs (u, v) of points in the handle itself and close to where the partition cuts the figure in two.

Another way to counteract imbalance is to use negative penalties to actively encourage splits between dissimilar points, rather than low positive penalties that merely discourage such splits less than the larger penalties associated with similar pairs do. For example, reducing C in

$$\sigma(u, v) := \sum_{q \in Q} \left(C - (q(v) - q(u))^2 \right),$$

which we used in Section 9.2 as a basis for (O5), has this effect, as does the use of $\langle u, v \rangle$ in (O7''). For every pair u, v split by a partition encoded as a function $f : V \to \{-1, 1\}$, such a negative penalty reduces its order. Hence if the partition divides many pairs u, v, as balanced partitions do, it will tend to have lower order than unbalanced partitions, as soon as the number of dissimilar pairs $u, v \in V$ (those with $\sigma(u, v) < 0$) substantially exceeds the number of similar pairs.

A downside of this approach is that the moment even a single penalty $\sigma(u, v)$ is negative, e.g. in (O5), we can lose the submodularity of the associated order function. This is different with the order function (O4) and its generalization

$$|f| = \tfrac{1}{2} f^\top K f \qquad\qquad (O6')$$

to advanced features. Recall that K is the $n \times n$ matrix, where $n = |V|$, whose entries in position (u, v) are $-\sigma(u, v)$ for $u \neq v$, and zero along the diagonal. Let K_c denote the matrix obtained from K by adding a positive constant c to every entry. Then

$$f^\top K_c f = f^\top K f + c \cdot f^\top(1) f,$$

where (1) denotes the $n \times n$ matrix with all entries 1. If $f: V \to \{-1, 1\}$ encodes a partition $\{A, B\}$ of V, then

$$f^\top(1)f = |A|^2 - 2|A||B| + |B|^2 = (|A| - |B|)^2.$$

This measures exactly what concerns us: the imbalance of the partition $\{A, B\}$. Now, as long as c is small enough that all the off-diagonal entries of K_c are still non-positive, the order function $f \mapsto |f|$ defined by

$$|f| := f^\top K_c f \qquad\qquad (O15)$$

will still be submodular [25]. But any positive c, no matter how small, discourages unbalanced partitions, by increasing their order more than it does for balanced partitions, when K_c replaces K. Thus, if our aim is to encourage balanced partitions while keeping our order function submodular, the prime contender is to use (O15) with c chosen as large as possible so that $c - \sigma(u, v) \leqslant 0$ for all distinct $u, v \in V$, i.e., setting

$$c := \min_{\{u,v\} \in V^{(2)}} \sigma(u, v) \geqslant 0.$$

9.9 Using order functions for deliberate bias

In this final section of this chapter we look at how to tweak order functions so as to encourage or discourage the consideration by our tangles of certain potential features that appear to be particularly relevant or irrelevant.

For example, we might be analysing a social survey with the specific aim of identifying musical tastes. If the survey was not compiled for this purpose, it will include many questions whose answers have little to do with what interests us. There is then a danger that these might obscure our results, by requiring our tangles to specify potential features that have nothing to do with musical tastes. Such potential features would seem irrelevant, and they should somehow be excluded from our analysis.

Conversely, we might wish that our tangles, even for small k, specify some potential features which we deem particularly important. Our discussions in Section 7.4 highlighted some problems with assigning orders to features explicitly. However, some of the generic order functions we discussed earlier in this chapter allow us to influence just a little bit what order they assign to some particular features we care about, while remaining generic and submodular.

Finally, there are the black hole tangles introduced in Section 2.4, which we may wish to ignore if they seem to have no meaningful interpretation.[36] Our choice of an order function can influence the occurrence of black hole tangles too, and we shall discuss this briefly at the end of this section.

Let us start, then, with the topic of how to weed out potentially irrelevant features from our feature system. The example indicated earlier was that we were trying to analyse a social survey with a specific aim to identify musical tastes. Since the survey was designed more broadly, many of its questions will be irrelevant to this task, and should perhaps be excluded from our analysis.

We have to be careful, though, in how we assess whether or not a feature is deemed 'irrelevant': if this is obvious from an interpretation point of view, as perhaps in the above example, then such features could simply be eliminated by hand. But can we be sure of which features to eliminate? For example, it may seem obvious that a question asking the participant's left- or right-handedness should be irrelevant to musical tastes. But what if it is not? Are we perhaps excluding the most interesting potential discoveries that tangles might unearth by removing the handedness question from our feature system?

It would help in such cases if there were an objective way to identify such potentially irrelevant features $\vec{s}$, by some criteria that take no
recourse to interpretation. One such criterion would be that our feature
system $\vec{S}$ has a pair of tangles that differ only in how they specify s. Let
us call such potential features s *local*. An example might be that s asks
a person's handedness, but handedness is irrelevant to taste in classical
music. Then any tangle of $S \smallsetminus \{s\}$ for liking Bach, for example, would
simply split into two tangles for liking Bach on S, a 'left-handed' and
a 'right-handed' one. Most of the other tangles of $S \smallsetminus \{s\}$ would split
similarly into two tangles of S, but s would distinguish no pair of tangles
that do not induce the same tangle of $S \smallsetminus \{s\}$.

Under most of the order functions we have considered in this chapter, such local features are assigned large order – which keeps them
out of the S_k for which we compute our tangles, those with k small.
However, having identified local features $\vec{s}$ mechanically, we can still
delete them from S explicitly (if this is desired after careful evaluation
of their interpretation) even if they have low order. Indeed, since s must
occur in any tree of tangles that distinguishes the pair of tangles which
only it distinguishes, deleting s from S will not jeopardize the existence
of the trees of tangles guaranteed for submodular feature systems by
Theorems 1 and 2, even if $S' := S \smallsetminus \{s\}$ is no longer submodular: if T is
a tree of tangles for S, then $T \smallsetminus \{s\}$ is a tree of tangles for S', with the
two tangles of S which only s distinguished merged into one tangle of S'.

Let us turn our attention now to the opposite of weeding out potentially irrelevant partitions, and see how we might encourage the adoption
into our feature system of some potential features that we deem particularly relevant. In Section 7.4 we discussed the option of simply defining
the orders of features explicitly, according to their perceived importance
in an application. We saw that this carried some disadvantages and
risks. Bias was among the risks, and a failure to be submodular was
among the disadvantages.

However it is possible to mimic explicit assignments of orders by
tweaking our generic order functions from Section 9.1, while keeping
them both generic and submodular. The idea is that, rather than assigning low order to important features explicitly, we can give them
greater weight in the definition of our similarity measure σ. Let us show
that this has a similar effect.

Let us choose a *weight function* $w \colon S \to \mathbb{N}$ that assigns large values
to important potential features. Our unbiased definition of σ then turns

into

$$\sigma_w(u, v) := \sum \{\, w(s) : s \in S, \ u(s) = v(s) \,\},$$

while our biased definition turns into

$$\sigma_w(u, v) := \sum \{\, w(s) : s \in S, \ u(s) = v(s) = \vec{s} \,\}.$$

We leave the definition of $|A, B|$ as $\sum_{a \in A} \sum_{b \in B} \sigma_w(a, b)$ unchanged. Similarly, our 'linear' order functions change from

$$\sum_{s \in S} \min(|A \cap \vec{s}|, |B \cap \vec{s}|) \quad \text{and} \quad \sum_{\vec{s} \in \vec{S}} \min(|A \cap \vec{s}|, |B \cap \vec{s}|)$$

for the biased and unbiased version to

$$\sum_{s \in S} w(s) \min(|A \cap \vec{s}|, |B \cap \vec{s}|) \quad \text{and} \quad \sum_{\vec{s} \in \vec{S}} w(s) \min(|A \cap \vec{s}|, |B \cap \vec{s}|),$$

respectively.

Note that increasing the weight of a potential feature $s = \{A, B\}$ does not affect the value of $|A, B|$ itself. Indeed, for the quadratic functions this is because $w(s)$ does not occur in the sum $\sigma_w(a, b)$ for any $a \in A$ and $b \in B$, as $a(s) \neq b(s)$ for such a and b. In the linear case, the minimum of $|A \cap A|$ and $|B \cap A|$, as well as the minimum of $|A \cap B|$ and $|B \cap B|$, is always zero, so s and $\vec{s}$ themselves account for zero in the sum for $|s|$ regardless of whether $\vec{s} = A$ or $\vec{s} = B$.

But raising the weight of s increases the order of most other $r \in S$: of every $r \in S \setminus \{s\}$ in the unbiased versions, and of every $r \in S$ that splits the set $\vec{s} \subseteq V$ in the biased versions. Hence if we increase $w(s)$, the order of r will rise: a little bit if r and s are similar as partitions of V, and a lot if the they are very different. Thus, indirectly, assigning s large weight amounts to assigning it, and elements of S similar to it, low order compared with the order of the other elements of S.

However, there is an important difference to choosing orders explicitly: order functions based on a weight function $s \mapsto w(s)$ in this way are defined on all the partitions $\{A, B\}$ of V, not just on S itself. And these order functions are submodular, with the same easy proofs as in the unweighted case.

Moreover, tangles of sets S_k defined with respect to such an order function are often guided by a weight function on V (see Section 2.5). Indeed, since our order function is submodular, the set $S_k^* \supseteq S_k$ of all the partitions of V of order $< k$ is submodular. It is known [19] that all the tangles of S_k^*, indeed of any submodular subset of S_k^*, are guided by weight functions on V. Hence any tangle of S_k that extends to a tangle of some submodular subset of S_k^* will also be guided by a weight function on V.

We thus have an indirect way of assigning low or high order to the potential features in S, explicitly if indirectly, which avoids the technical problems that this would cause if we did it directly. When we later say in our discussion of applications that we might choose the order of some features explicitly in one way or another, then doing it in this indirect way will often be what is meant.

Finally, let us see how our choice of an order function can influence the existence of black hole tangles as discussed in Section 2.4. We shall illustrate this by the furniture example from Section 2.2, in which $S = \{p, q, r, s\}$ and $A = \vec{p}$, $B = \vec{q}$, $C = \vec{r}$ and $D = \vec{s}$ stood for the properties of being made of wood, steel, wicker or plastic. The specification $\{\overleftarrow{p}, \overleftarrow{q}, \overleftarrow{r}, \overleftarrow{s}\}$ of S itself is a black hole tangle of S.

This is not a particularly interesting tangle from an application point of view: it just signifies the absence of any dominating material in the composition of the furniture considered. But then, in a real application we would likely not try to compute tangles of this S directly, but of suitable S_k^* in the set S^* of all the partitions of V. So let us consider these, with respect to an order function determined by a similarity measure σ based on our original S, as defined in Section 9.1.

With the unbiased definition of σ, the partitions in S_1^*, those of order 0, are precisely the ones in S: they are the partitions pitching one material against the other three. In particular, partitions such as $\{A \cup B, C \cup D\}$ are not in S_1^*, because they cut right through features such as $A \cup B \cup C = \overleftarrow{s}$ and thus have large order. Hence for the unbiased definition of σ, nothing much has changed: we still get that undesirable void black hole tangle for S_1^*.

When we base our order function on the biased version of σ, however, with default specifications A, B, C, D, this is no longer a tangle. The reason is that S_1^* has now grown to include partitions of V that pitch two materials against the other two, such as $\{A \cup B, C \cup D\}$: the fact that this cuts right through the *inverse* $A \cup B \cup C$ of the feature D is no longer penalized by σ. So any tangle of S_1^* has to orient these new partitions too. But a moment's thought shows that the only consistent orientations of this larger set S_1^* are the four desired ones, those that orient all the partitions towards one of the four materials.[37]

10

Choosing the feature system

The tangle theory presented so far always started with a given feature system $\vec{S}$, or perhaps a hierarchy $\vec{S_1} \subseteq \vec{S_2} \subseteq \ldots$ of feature systems. When we apply tangles in real-world scenarios, however, we are usually free to choose these feature systems – and tasked with doing so in a way that optimally suits the application. This chapter is devoted to the question of how to do this well.

Regardless of whether we end up considering tangles of a single feature system $\vec{S}$ or of a hierarchy $\vec{S_1} \subseteq \vec{S_2} \subseteq \ldots$ of feature systems, there are some fundamental questions that concern both cases. Put another way, even a hierarchy $\vec{S_1} \subseteq \vec{S_2} \subseteq \ldots$ of feature systems lives inside one big system $\vec{S}$, which has to be chosen first. Once that is done we can choose a hierarchy inside it if we wish, which we usually do by way of picking an order function on S as discussed in Chapter 9.

There are three main ways in which S can arise naturally in the context of the tangle applications discussed in this book:

(i) as a content-specific set of measurements, questions in a survey, natural partitions of a given dataset (e.g., lines across an image);

(ii) as the set of all the partitions of a dataset V, or a random sample generated from this set;

(iii) as a set of partitions of V computed by some standard method in data science, such as neural networks or spectral techniques.

Often we extend a set S as above by adding infima and suprema not in $\vec{S}$ of features already in $\vec{S}$, to make $\vec{S}$ submodular or locally submodular as our tree-of-tangles algorithm requires it. In addition, for every feature system there is a dual system as defined in Section 7.7, which may be less obvious but equally useful.

In the case of (i) above there is nothing further for us to choose, so this case will not be treated in this chapter. We shall look briefly into case (ii), and then devote two sections to special instances of (iii).

10.1 Random partitions, locally optimized

The simplest paradigm for constructing S from randomly chosen partitions of V is that we generate one such partition at a time, optimize it by moving elements of V across to reduce its order if possible, and when this process stops adopt the partition then current into our set S.

As indicated in Section 9.8, it will usually be desirable to start this process with a partition of V that is reasonably balanced: a partition that has roughly half of V on either side. This will automatically happen with high probability if we choose that starting partition at random, since there are more balanced partitions than unbalanced ones. However we can additionally encourage the construction of balanced partitions in the iterative process that follows, by tweaking our order function as indicated in Section 9.8.

In this iterative process we look at each element of V, or perhaps at groups of such elements on the same side of the current partition that are pairwise similar in terms of some similarity function σ, and then move such elements across the partition in a bid to reduce its order. Whether this process is more likely to end with a balanced or an unbalanced partition therefore depends on the order function we use here.

Let us briefly indicate another way to generate S. It also starts from a random partition $\{A, B\}$ of V and proceeds iteratively, but not by moving elements of V across the partition. In the interest of brevity, the following description of this approach uses graph terminology; any reader unfamiliar with this can skip the next paragraph without much loss.

The idea is to find, in the complete graph on V with edge weights $\sigma(u, v)$ or $\langle u, v \rangle$ as discussed in Chapter 9, a minimum cut between sets $A' \subseteq A$ and $B' \subseteq B$ on different sides of the current partition $\{A, B\}$.[1] These sets A' and B' should not be too small, to avoid generating an unbalanced cut. They could be chosen at random inside A and B, they could be sets of pairwise similar elements inducing dense subgraphs, or they could be balls of some fixed radius around representatives $a \in A$ and $b \in B$. Note that the current partition $\{A, B\}$ is one possible solution to this optimization problem. But $\{A, B\}$ may be replaced with a cut of smaller value, a partition of lower order, that still separates A' from B'. If this does not happen for a few choices of A' and B', we stop the iteration and adopt $\{A, B\}$ for our collection S.

10.2 Spectral feature systems

The so-called *spectral*[2] method allows us to compute a set S of partitions of V that is smaller than the set of all partitions, but more meaningful than a random set of partitions of the same size can be expected to be.

Before we delve into the formal details of how to construct such a set S, let me indicate in a brief high-level description what is going to happen here, what this approach has in common with standard spectral clustering [29], and how it differs from it.[3] Readers unfamiliar with spectral clustering are invited to simply skip the next two paragraphs.

As in spectral clustering, the basic ingredients from which we build S are some eigenvectors of the Laplacian L defined in Section 9.2. More precisely, we shall work with orthogonal eigenvectors of the ℓ smallest eigenvalues (listed with multiplicities, i.e. not necessarily distinct), for some $\ell \leqslant |V|$ that we are free to choose. Spectral clustering now embeds V in an ℓ-dimensional space defined by those ℓ eigenvectors, runs a standard clustering algorithm such as 'k-means' on the embedded points, and finishes by translating the clusters found there back into clusters of V.

Our approach instead translates each of those eigenvectors into a partition of V to be put in S. We then apply tangles to this S, or to a suitable hierarchy $S_1, S_2, \ldots$ defined by some order function as discussed in Chapter 9. So we basically divert from spectral clustering only once it has done its spectral magic and moves on to ordinary clustering. There, our own work begins: rather than perform clustering on some spectrally defined image of V in $\mathbb{R}^\ell$ we analyse the tangles of this S.

Let us now describe the process of generating S 'by spectral methods' in more detail. We assume that we have a similarity measure $\sigma\colon V^{(2)} \to \mathbb{R}$ on V that takes non-negative values throughout. As usual, we abbreviate $\sigma(\{u, v\})$ to $\sigma(u, v) = \sigma(v, u)$. Let L be the Laplacian associated with σ, as defined in Section 9.2: this is a square matrix whose rows and columns are each indexed by V. Its off-diagonal entries are $-\sigma(u, v)$, its diagonal entries $\sigma(u, u)$ are $\sum_{v \neq u} \sigma(u, v)$. In order to speed up our computations, such as of the eigenvalues of L, we will often round small values of $\sigma(u, v)$ to zero.

The matrix L has $n = |V|$ real eigenvalues $0 = \lambda_1 \leqslant \ldots \leqslant \lambda_n$. These λ_i have eigenvectors $f_i \colon V \to \mathbb{R}$ such that $\langle f_i, f_j \rangle = 0$ for all $i \neq j$. This latter condition has a name: the f_i are pairwise *orthogonal*. As f_1 we can, and usually do, choose the function $\mathbb{1}$ that sends every $v \in V$ to 1. For $i \geqslant 2$ we can, and will, choose the f_i so that $\langle f_i, f_i \rangle = 1$. We refer to this fact by saying that the f_i have been *normalized*.

As in Section 9.2 we often encode the functions $f \colon V \to \mathbb{R}$ as $n \times 1$ matrices, indexed by V, whose entry in row v is $f(v)$. Then f is normalized if and only if $f^\top f = 1$. Recall that L defines a submodular order function on all f via

$$|f| = f^\top L f \geqslant 0. \tag{O5$'$}$$

Our special eigenvectors $f_2, \ldots f_n$ satisfy this with $|f_i| = f_i^\top L f_i = \lambda_i$.

Our aim is to generate a set S of partitions of V, or more generally a set $\vec{S}$ of advanced features $f \colon V \to \mathbb{R}$, whose tangles we intend to compute. We first generate some advanced features $f \colon V \to \mathbb{R}$ for $\vec{S}$, rather than partitions of V, even if at the end of the day we are interested in partitions. We can turn these advanced features f into partitions at a later stage if desired, and will discuss how this can best be done towards the end of this section.

We shall assume in what follows that V has no partition $\{A, B\}$ such that $\sigma(a, b) = 0$ for all $a \in A$ and $b \in B$. If there is such a partition, we add it to S straight away (in other words: we add $f \colon V \to \mathbb{R}$ to $\vec{S}$ that sends A to -1 and B to 1) and start again with V replaced by A and, separately, by B.[4]

Under this assumption, the first eigenvalue $\lambda_1 = 0$ of L has multiplicity 1. This means that $0 < \lambda_2 \leqslant \ldots \leqslant \lambda_n$. We now pick eigenvectors $f_2, f_3, \ldots$ for the eigenvalues $\lambda_2, \lambda_3, \ldots$ as features for $\vec{S}$, one by one, all orthogonal to both $f_1 = \mathbb{1}$ and to each other. We could do this all the way up to $i = n$. But we usually stop at some earlier $\ell \leqslant n$, once we have what looks like a good supply of features for our system $\vec{S}$ to ensure it has interesting tangles.

Any sequence $f_2, \ldots, f_\ell$ picked in this way has the following properties. Since f_i is orthogonal to f_1, its values sum to zero: $\sum_{v \in V} f_i(v) = 0$. Since f_i is not all-zero, this means that it has positive as well as negative values. If desired, we could turn it into a partition of V based on this fact alone, by replacing all its positive values with 1 and its negative values with -1, breaking ties of $f_i(v) = 0$ arbitrarily. This modified function would no longer be an eigenvector of λ_i. But if the eigenvalues and eigenvectors have worked their magic and all we want to keep is a set of partitions of V, then this is a simple way to do that.

A more fundamental property of the f_i is that each of them minimizes

$$|f| = f^\top L f$$

amongst all the normalized $f \colon V \to \mathbb{R}$ that are orthogonal to $f_1, \ldots, f_{i-1}$.

In fact, more is true: the entire collection $f_1, \ldots, f_n$ minimizes a similar objective function globally. To state this precisely in its simplest form, let us normalize also f_1 by giving it values $f_1(v) = 1/\sqrt{n}$ rather than 1 for all $v \in V$. Now consider *any* collection $f_1', \ldots, f_\ell'$ of normalized $V \to \mathbb{R}$ functions that are pairwise orthogonal, with $\ell \leqslant n$. Let us write these as the columns of an $n \times \ell$ matrix H, and consider the $\ell \times \ell$ matrix $H^\top H$. This has entries $\langle f_i', f_i' \rangle = 1$ along the diagonal, and all other entries $\langle f_i', f_j' \rangle = 0$: it is the *identity matrix* I_ℓ. Let $\mathcal{H}$ be the set of all $n \times \ell$ matrices H obtained in this way.

Given $H \in \mathcal{H}$, we can also compute the matrix $H^\top L H$: this is another $\ell \times \ell$ matrix. The *trace* of this matrix is the sum of its diagonal entries, denoted as $\mathrm{tr}(H^\top L H)$. Note that, for $\ell = 1$ and f the sole column of H, this is exactly our earlier $|f| = f^\top L f$. In general, we have

$$\mathrm{tr}(H^\top L H) = \sum_{i=1}^{\ell} {f_i'}^\top L f_i' = |f_1'| + \ldots + |f_\ell'|.$$

Linear algebra [23, Cor. 9.5.2] tells us that this sum is minimum, compared with all other $H \in \mathcal{H}$, when H is built from our chosen eigenvectors $f_1, \ldots, f_\ell$ as its columns. So not only does each f_i have minimum order when compared with other normalized f orthogonal to $f_1, \ldots, f_{i-1}$, but also the entire sum of all their orders $|f_i|$ is minimum when compared with all possible families of normalized orthogonal functions $f_1', \ldots, f_\ell'$.

In other words: these f_i are bottlenecks of V in quite a strong sense. They have low order and, when viewed as partitions of V as indicated above, they divide it in a reasonably balanced way: the sum of their negative values equals, in absolute terms, the sum of their positive values.

Let us return now to the question of how to translate the advanced features $f_1, \ldots, f_\ell$ we have collected for $\vec{S}$ into partitions of V, if that is indeed our aim. If none of the f_i maps any v to 0, there is a standard way to do this: since each f_i has both positive and negative values, it partitions V into those v with $f_i(v) > 0$ and those with $f_i(v) < 0$. We orient this partition towards the set with positive values. If 0 occurs as a value, we let each f_i partition V into three rather than two parts, by replacing its positive values with 1, its negative values with -1, and leaving its zero values untouched. The triples defined by these simplified advanced features are known as set separations (rather than set partitions); their tangles have been studied at least as well as those of set partitions.

Another option is to use some more creative way to divide the set $f_i(V) \subseteq \mathbb{R}$ of values of f_i into subsets C_i^j, with $j = 1, \ldots, k(i)$ say, each of which defines a partition of V into $\{ v \mid f_i(v) \in C_i^j \}$ and its complement. For example, we might split $f_i(V)$ into two sets as earlier, but at some other natural threshold than its average (which our earlier $+/-$ split does, because $\langle 1, f_i \rangle = 0$), such as its median, or at its largest gap. Or we might use some standard clustering algorithm on the set $f_i(V) \subseteq \mathbb{R}$, such as k-means, in which case k becomes the $k(i)$ above. That could amount to a little instance of spectral clustering in its own right [29] – except that we do not stop there, but use the partitions obtained in this way as our set S whose tangles we then compute.

Another way to convert each f_i into several oriented partitions of V is to compute its 'nodal domains'. In graph-theoretic language, these are defined as follows. Let G be the *similarity graph* on V defined by σ, the graph with edges $\{u, v\}$ whenever $\sigma(u, v) > 0$. For each f_i consider the sets $A_i = \{ v \mid f_i(v) \leqslant 0 \}$ and $B_i = \{ v \mid f_i(v) \geqslant 0 \}$. The *nodal domains* of f_i are either the components of $G[A_i]$ and of $G[B_i]$, or those of $G[A_i \smallsetminus B_i]$ and of $G[B_i \smallsetminus A_i]$; both versions exist. The partitions corresponding to these nodal domains of f_i are those that pitch one of them against the rest of V, separately for each i, as for our earlier sets $f_i^{-1}(C_i^j)$.

As indicated at the start of this chapter, once we have collected enough features $f_2, f_3, \ldots$ into our system $\vec{S}$, we can either compute its tangles straight away or, depending on the application, enrich $\vec{S}$ by adding some infima and suprema in the process of constructing their tree of tangles, as required by our algorithm in Section 11.3.

Let us close by mentioning another way to utilize our spectrally determined features $f_1, \ldots, f_\ell$, or any partitions of V they define. If we want $\vec{S}$ to be an advanced feature system, let it consist of all functions $f : V \to \mathbb{R}$, with the same range such as $\{-1, 1\}$ or $\{-5, \ldots, 5\}$ as before, that satisfy $f(u) = f(v)$ whenever $f_i(u) = f_i(v)$ for all $i = 1, \ldots, \ell$, or even if, for all i, the values $f_i(u)$ and $f_i(v)$ lie in some common traditional 1-dimensional 'cluster' of the range of f_i in $\mathbb{R}$, as earlier. Note that these are far fewer functions than all $V \to \mathbb{R}$ functions with this range.

If we want $\vec{S}$ to be an ordinary feature system of partitions of V, we first turn our f_i into oriented partitions $\vec{s_i}$ of V, as described earlier. We then let S consists of all the partitions of V that split only points $u, v \in V$ that are also split by at least one f_i: groups of v on which all the f_i agree, or nearly agree in that they place u and v in a common traditional 1-dimensional 'cluster' of the range of f_i, must not be split by S, but all partitions of V that observe this are in S. Once more, these are far fewer partitions of V than all.

We have thus reduced the set of all partitions of V, or more generally of all the $V \to \mathbb{R}$ functions with our desired range, to a much smaller set $\vec{S}$. Since $f_1, \dots, f_\ell$ were chosen of particularly low order, the $V \to \mathbb{R}$ functions or partitions of V in our reduced set $\vec{S}$ are likely to be 'good' already. But this need not stop us from defining inside it a hierarchy $\vec{S_1} \subseteq \vec{S_2} \dots$ of features of even lower order.

This can be done with respect to any of the order functions we discussed in Chapter 9. In particular, we can re-use the similarity measure $\sigma \colon V^{(2)} \to \mathbb{R}$ on which we built the order $|f| = f^{\top}Lf$ we used to find $f_1, \dots, f_\ell$, or even re-use that order function itself on our new smaller $\vec{S}$ to define a hierarchy $\vec{S_1} \subseteq \vec{S_2} \dots$ inside it.

10.3 PCA-related feature systems

In Section 10.2 we used the Laplacian L to find some bottlenecks, advanced features $f_2, \dots, f_\ell$ mapping V to $\mathbb{R}$, or partitions of V induced by them, that we could use as building blocks for a feature system $\vec{S}$. Our aim was that the tangles of S should describe the structure imposed on V by a similarity measure $\sigma \colon V^{(2)} \to \mathbb{R}$. This σ would typically be defined by how the 'people' in V answered a set Q of 'questions', viewed as maps $q \colon V \to \{-5, \dots, 5\}$. The reason this can work well in practice is that the features $f_2, \dots, f_\ell$ found in this way, as orthogonal eigenvectors of L, had some desirable properties:

- they had low order $|f| = f^{\top}Lf$, as defined in (O5');
- they were more balanced than other low-order features for (O5);
- they could be computed fast if L was sparse, which however required that we round small similarities $\sigma(u, v)$ to zero.

Our aim in this section is to follow a similar approach based not on the Laplacian L but on the data matrix J introduced in Section 9.3, which directly records how the people in V answered the questions in Q. The features $f_1, \ldots, f_\ell$ we find in this way will have similar advantages:

- they will have low order $|f| = f^\top J f$, as defined in (O7′);
- they reflect the structure of our data directly, not via similarities, forming the first ℓ 'principal components' of our data;
- they can be computed from eigenvectors of a much smaller matrix.

Recall that J is a symmetric $n \times n$ matrix defined as

$$J := -MM^\top,$$

where M is the $n \times m$ matrix whose $n = |V|$ rows are indexed by the elements of V, whose $m = |Q|$ columns are indexed by the elements of Q, and which has entries $q(v)$ in row v and column q. The matrix J has entries

$$-\langle u, v \rangle = -\sum_{q \in Q} q(u) \cdot q(v)$$

in row u and column v; note that $-\langle u, v \rangle$ is defined, and equal to the sum on the right, if we interpret u and v as $Q \to \mathbb{R}$ functions $q \mapsto q(u)$ and $q \mapsto q(v)$, so that $u(q) = q(u)$ and $v(q) = q(v)$.

In (O7′) we used J to assign an order $|f|$ to every $f: V \to \mathbb{R}$, given by

$$|f| = f^\top J f = -\sum_{q \in Q} \langle f, q \rangle^2 \leqslant 0. \tag{O7}$$

As $f^\top J f \leqslant 0$ for all f, J has n real eigenvalues $\lambda_1 \leqslant \ldots \leqslant \lambda_n \leqslant 0$.

Just as in Section 10.2 we pick eigenvectors f_i for these λ_i in turn, choosing each f_i normalized and orthogonal to $f_1, \ldots, f_{i-1}$. This alone ensures that, for all $\ell \leqslant n$, the sum $|f_1| + \ldots + |f_\ell|$ is minimum amongst all sums $|f_1'| + \ldots + |f_\ell'|$ such that $\langle f_i', f_i' \rangle = 1$ for all i and $\langle f_i', f_j' \rangle = 0$ for all $i < j$, by [23, Cor. 9.5.2]. We can then choose $\ell \leqslant n$ as we wish, and build $\vec{S}$ from $f_1, \ldots, f_\ell$ as we did in Section 10.2: either just adding negatives, infima and suprema iteratively as required for our tree of tangles, or as the set of all functions $f: V \to \mathbb{R}$ with the same range as for Q that assign the same value to any $u, v \in V$ such that $f_i(u)$ and $f_i(v)$ either coincide or at least lie in a common traditional 1-dimensional 'cluster' of the range of f_i in $\mathbb{R}$, for all $i = 1, \ldots, \ell$.

These eigenvectors could be easily computed in the case of the Laplacian L when this was sparse, which can often be achieved by rounding $\sigma(u, v)$ to zero when it is small. Our matrix J, however, will not normally be sparse, and so computing its eigenvectors directly may be impractical when n is large. However, since J has the special form of $J = -MM^\top$, we shall be able to derive all its relevant eigenvectors from those of the usually much smaller $m \times m$ matrix

$$I := -M^\top M.$$

Indeed, for any real $\lambda \neq 0$, if u is an eigenvector of I for λ then Mu is easily seen to be an eigenvector of J for λ, and if f is an eigenvector of J for λ then $M^\top f$ is an eigenvector of I for λ. These maps $u \mapsto Mu$ and $f \mapsto M^\top f$ are inverse to each other up to a constant: $M^\top Mu = -Iu = -\lambda u$, and $MM^\top f = -Jf = -\lambda f$. In particular, I and J have the same non-zero eigenvalues with the same multiplicities.

We can thus derive our desired eigenvectors $f_1, \ldots, f_\ell$ of J from the corresponding eigenvectors of I, as long as the eigenvalues $\lambda_1, \ldots, \lambda_\ell$ are non-zero. But since $\lambda_1 \leqslant \ldots \leqslant \lambda_n \leqslant 0$, this will rarely be a problem: we just stop collecting $f_1, \ldots, f_\ell$ at an ℓ small enough that $\lambda_\ell \neq 0$.

In any case we have $\ell \leqslant m$, since I is an $m \times m$ matrix. Now m, the number of questions, is usually much smaller than the number n of people that answered them. And if Q contains groups of similar questions, we are likely to choose at most one eigenvector f_i for every such group.

The matrix I and its eigenvectors are also interesting in their own right, and they throw a new light on J. We use the rest of this section to discuss this. We shall also see that, under some natural normalization assumptions, the u_i are the so-called 'principal components' of the $v \in V$ viewed as $Q \to \mathbb{R}$ functions, while the f_i are the 'principal components' of the questions $q \in Q$ viewed as $V \to \mathbb{R}$ functions (after normalization).

By definition, I is a symmetric $m \times m$ matrix. We think of it as indexed by the elements of Q, just as J is indexed by the elements of V. It has entries

$$-\langle p, q \rangle = -\sum_{v \in V} v(p) \cdot v(q)$$

in row p and column q, where $v \colon Q \to \mathbb{R}$ satisfies $v(q) = q(v)$ for all $q \in Q$.

Dually to (O7), (O7′) and (O7″), we can define an 'order'

$$|u| := u^\top Iu = -\sum_{(p,q) \in Q^2} u(p) \langle p, q \rangle u(q) = -\sum_{v \in V} \langle u, v \rangle^2 \leqslant 0 \qquad (\overline{\text{O}}7)$$

on the functions $u \colon Q \to \mathbb{R}$, which in matrix multiplications we interpret as $m \times 1$ matrices indexed by Q, with entry $u(q)$ in position q.[5]

As $u^\top I u \leqslant 0$ for all u, I has m real eigenvalues $\lambda_1 \leqslant \ldots \leqslant \lambda_m \leqslant 0$. We already know that all these eigenvalues, except possibly any zero eigenvalues at the end, coincide with the corresponding eigenvalues of J.

Dually to our earlier observations about the eigenvectors of J for these eigenvalues, and by the same fact from linear algebra [23], any choice of eigenvectors $u_1, \ldots, u_m$ of I for $\lambda_1 \leqslant \ldots \leqslant \lambda_m$ such that each u_i is normalized and orthogonal to $u_1, \ldots, u_{i-1}$ has the property that, for every $\ell \leqslant m$, the sum $|u_1| + \ldots + |u_\ell|$ is minimum amongst all sums $|u_1'| + \ldots + |u_\ell'|$ such that $\langle u_i', u_i' \rangle = 1$ for all i and $\langle u_i', u_j' \rangle = 0$ for all $i < j$. In particular, by $(\overline{\text{O}}7)$, choosing u as u_i

- maximizes $\sum_{v \in V} \langle u, v \rangle^2$ over all normalized u orthogonal to all of $u_1, \ldots, u_{i-1}$, for all i.

In plain words this says that, amongst all the u eligible for its role, our u_i is most strongly correlated, positively or negatively, to the various $v \in V$ when we compare them as $Q \to \mathbb{R}$ functions. Informally, we may think of u_i as having the potential to reflect the most 'additional information' about our dataset V given the information already reflected by $u_1, \ldots, u_{i-1}$. See our discussion of (O7) in Section 9.3.

There is another interesting interpretation of this if we make one additional assumption: that for every question in Q the numerical answers it received from the various $v \in V$ have average zero. This is not an unreasonable assumption: it simply balances out any intrinsic positive or negative slant in the phrasing of q. We can achieve it for each q retrospectively by shifting the map $q: V \to \mathbb{R}$, adding a suitable constant to achieve $\sum_{v \in V} v(q) = 0$.

Let us call such q *balanced*, and assume from now on that all the questions in Q are balanced. Under this assumption, every $u: Q \to \mathbb{R}$ we considered for u_i satisfies

$$\sum_{v \in V} \langle u, v \rangle = \sum_{v \in V} \sum_{q \in Q} u(q) v(q) = \sum_{q \in Q} u(q) \sum_{v \in V} v(q) = 0.$$

Hence, for every fixed u, the $V \to \mathbb{R}$ function $v \mapsto \langle u, v \rangle$ is balanced, too.

Geometrically, the number $|\langle u, v \rangle|$ is the length of the orthogonal projection $\langle u, v \rangle u$ of $v \in \mathbb{R}^Q$ to the 1-dimensional subspace $\mathbb{R}u$ of $\mathbb{R}^Q$ spanned by our normalized $u \in \mathbb{R}^Q$. As noted earlier, our choice of u_i maximizes, over all eligible u, the sum – or equivalently, the average – of the squares $\langle u, v \rangle^2$ of these numbers for all $v \in V$.

Given any balanced $V \to \mathbb{R}$ function, the average of the squares of its values is known as its *variance*. It measures how widely spread these values are as v varies over V. Viewed in these terms, our choice of u_i thus also

- maximizes the variance of the lengths of the projections to $\mathbb{R}u$ of the $v \in V$, over all normalized u orthogonal to all of $u_1, \ldots, u_{i-1}$.

The $Q \to \mathbb{R}$ functions $u_1, \ldots, u_\ell$ are known as the first ℓ *principal components* of our $Q \to \mathbb{R}$ functions $v \in V$, which represent all our data. Similarly, if we normalize the answers to Q given by each $v \in V$ to make the functions $v \colon Q \to \mathbb{R}$ balanced, then the $Q \to \mathbb{R}$ functions $q \to \langle f, q \rangle$ will be balanced too, for every fixed $f \colon V \to \mathbb{R}$. Our special such functions $f_1, f_2, \ldots$, constructed earlier as eigenvectors for J, are then the principal components for the questions in Q. Their advanced-feature tangles, then, are selections of some of these principal components, choosing f_i for some i and $-f_i$ for the others.[6]

The above maximality condition for the choice of u_i can be rewritten as follows. Every point $v \in \mathbb{R}^Q \cong \mathbb{R}^m$ forms a right-angled triangle with its projection $\langle u, v \rangle u$ to $\mathbb{R}u$ and the origin. The two shorter sides of this triangle have lengths $|\langle u, v \rangle|$ and $d(v, \mathbb{R}u)$, respectively, where $d(v, \mathbb{R}u)$ denotes the distance between v the point nearest to it on the line $\mathbb{R}u$. The third side, the hypotenuse, is the line segment between the origin and v. Now if we consider v as fixed and let u vary, then the hypotenuse has some fixed length. By Pythagoras's theorem the squares of the lengths of the other two sides, $\langle u, v \rangle^2$ and $d(v, \mathbb{R}u)^2$, sum to the square of the hypothenuse, some fixed number. Hence maximizing $\sum_{v \in V} \langle u, v \rangle^2$, as our condition for u_i does, is equivalent to minimizing $\sum_{v \in V} d(v, \mathbb{R}u)^2$. We can thus rephrase our condition for the choice of u_i to say that it

- minimizes the average squared distance of the $v \in V$ from $\mathbb{R}u$, over all normalized u orthogonal to all of $u_1, \ldots, u_{i-1}$.

This condition has yet another interpretation. To see this, let us view every $u \colon Q \to \mathbb{R}$ as an m-tuple of coefficients $u(q)$, one for each $q \in Q$, in the linear combination

$$q_u := \sum_{q \in Q} u(q)q = Mu.$$

Like all the $q \in Q$ from which it is combined, this q_u is a balanced $V \to \mathbb{R}$

function, with values

$$q_u(v) = \sum_{q \in Q} u(q)q(v) = \langle u, v \rangle$$

for every $v \in V$. When Q consists of questions, we may think of q_u as a combined question, already answered by all the $v \in V$, in which each $q \in Q$ is weighted according to how u answered it. As M has these q as its columns, we then have

$$|u| = u^\mathsf{T} I u = -u^\mathsf{T} M^\mathsf{T} M u = -\langle Mu, Mu \rangle = -\langle q_u, q_u \rangle = -\|q_u\|^2,$$

where $\|q_u\|$ is the length of $q_u \colon V \to \mathbb{R}$ viewed as a vector in $\mathbb{R}^n$.

Our choice of u_i thus also maximizes $\langle q_u, q_u \rangle$ amongst all the u eligible to be chosen as u_i. As $|V|$ is constant, this is tantamount to maximizing the variance $\sum_{v \in V} q_u^2(v)/|V| = \langle q_u, q_u \rangle/|V|$ of q_u. Hence, u_i also

- maximizes the variance of q_u over all normalized u orthogonal to all of $u_1, \ldots, u_{i-1}$.

Recall that, as u_i is an eigenvector of I for λ_i, the $V \to \mathbb{R}$ map $q_{u_i} = Mu_i$ is an eigenvector of J for λ_i. By re-ordering the eigenvectors u_i or f_i for a given eigenvalue λ, we may thus assume that $f_i = q_{u_i}$ for all i such that $\lambda_i \neq 0$.

11

Algorithms

There are a number of algorithmic aspects concerning tangles. These include the computation of the following objects:

- the set S, or sets S_k, in whose tangles we are interested;
- the tangles of S or of the S_k;
- if tangles are found, a tree set T of potential features distinguishing them (as in Theorems 1 or 2);
- if no tangles are found, a tree set T of potential features that certifies their non-existence (as in Theorem 3).

The four sections in this chapter address these points in turn. Software that implements our algorithms, and more sophisticated alternatives, is available at [1].

11.1 Generating the feature systems

In Chapter 10 we already discussed in some detail how to choose the feature system $\vec{S}$ that should underly our tangles. In this section we look at this question again, but more from a computational angle. In particular, when our theoretical set of potential features is very large, such as the set of all partitions of some large dataset V, our task will be to choose S not only so that it captures the structure of our data optimally, but also in such a way that we can compute in practice either all its tangles or at least all its k-tangles for small k.

We shall discuss this question separately for our two basic scenarios: the questionnaire scenario and the partition scenario.

Let us assume first that we are in a questionnaire-type scenario. There we start from a set Q of questions that have been answered by the people in V. In the context of Section 1.1 these 'questions' might be measurements taken of some phenomena $v \in V$ we observed, but for simplicity let us stick to the terms of 'questions' and 'people' as in Section 1.2. Our task is to extract from this a set S of potential features – or perhaps a hierarchy of sets S_k for small k only, if we have settled for an order function defined on Q. This begins with the task of turning the questions from Q into a format we can work with.

The desired format could be that of functions $V \to \mathbb{R}$ as discussed in Section 7.6. Or we might wish to turn the questions from Q into yes/no questions, even if they come as $V \to \mathbb{R}$ functions whose tangles we could compute directly. This conversion may lead to different tangles depending on how we do it, and it is not clear from the outset which of these are more informative. So it may be a good idea to experiment with all the options, even on the same dataset.

If we wish to work with a biased order function, as discussed in Section 9.1, we have to prepare our data by fixing default specifications $\vec{s}$ for the potential features s to be adopted into S or S_k. These default specifications are likely to be given naturally for the questions in Q, as in the shopping cart tangles in Section 6.3, but for any $s \in S \setminus Q$ we shall still have to specify them.

It is almost always a good idea not to compute all the tangles of a set S obtained from Q as above, but to compute its k-tangles for increasing k with respect to some well-chosen order function. As discussed in Chapter 9, this can help us weed out irrelevant[1] questions from a questionnaire, because they are assigned large orders.

Once we have chosen a good order function, however, the tangles we shall be interested in are definitely the k-tangles of our fixed set S obtained from Q in some controlled way: unlike in the case when S is chosen blindly from the set of all partitions of some dataset V, the partition scenario to be discussed later, our set S in a questionnaire scenario will still consist of questions that can be interpreted, and we are interested in the typical sets of answers to precisely these questions.

This difference matters when we enlarge S by adding corners of features, as we typically do when we compute our tree of tangles. While corners of arbitrary partitions of a set V are just other partitions, no worse than the original ones, corners added in a questionnaire context will correspond to Boolean expressions of our original questions from S.

They can still be interpreted in principle – but when the expression is complicated, the interpretation sometimes makes little intuitive sense.

Another aspect is that adding low-order corners enlarges our set S_k, and these larger sets may have fewer tangles than the original S_k. As we shall see below, the tangles of S_k that do not extend to this larger set are always bad, or 'fake', and we shall be glad to have spotted and excised them. In our questionnaire scenario, however, the opposite is true too: we are primarily interested in the tangles, or k-tangles, of Q; whether or not they extend to tangles of larger sets obtained by adding corners will be interesting to know but may not be a sine qua non.

Now assume that we are in a partition scenario, where in principle we consider the entire set of all $2^{|V|}$ partitions of V. Let us assume that we have settled for an order function on these partitions, on the basis of our considerations in Chapter 9. Our algorithmic task now is to generate, from the set of all the partitions of V, a set S of partitions whose k-tangles we will then compute. Our approach is to generate the sets S_k in turn for increasing k, and simply call their union S at the end. In Chapter 10 we discussed a few ways of choosing such sets S_k, but there are still computational details to be decided.

One is how to come up with a starting partition s for our iterative process in which we reduce the order of s until we hit a local minimum, at which point we either adopt the then current s into the relevant sets S_k or discard it and start again. In Section 10.1 we chose this starting s 'at random', and computationally there are many ways of doing that.

We might try to be just a little cleverer than random and start with partitions s obtained by some standard partitioning algorithm that divides V into two parts. Partitions returned by such an algorithm might well be local minima already with respect to moving elements of V across.

In Sections 10.2 and 10.3 we obtained these s from eigenvectors of certain matrices. Again, there are several ways of turning an eigenvector $f : V \to \mathbb{R}$ into a partition of V: we could divide V into $\{ v \in V \mid f(v) < 0 \}$ and $\{ v \in V \mid f(v) \geqslant 0 \}$, but we could also look for natural 1-dimensional clusters amongst the values $f(v)$ in $\mathbb{R}$ and divide V so as not to cut right through any of those clusters.

More fundamentally, the spectral techniques from Sections 10.2 and 10.3 offered us not just one starting partition s of V but a sequence of partitions, adapted from a sequence of orthogonal eigenvectors. We may adopt such an entire sequence as starting partitions s for our iterative procedure. But we also have the option of taking just a few, perhaps just the first of these. If this one partition is $\{A, B\}$, say, we might then proceed by looking to partition A and B further (rather than starting over with a new partition of all of V), perhaps again using spectral techniques. The difference here is that the partitions obtained recursively in this way would not have to come from orthogonal eigenvectors of the same matrix – which may be a blessing or a curse and has to be tested.

In general we prefer to avoid such iterative, or 'hierarchical', ways of choosing our partitions of V.[2] Such iterative processes produce nested partitions: first $\{A, B\}$, say, then complementary subsets of A versus one another (add B to both to obtain partitions of V), and so on. But tangles are at their best when the partitions they orient cross a lot. Highly crossing partition sets S do not define standard hierarchical clusters, as nested sets do, but tangles can still capture 'fuzzy' clusters in V by orienting those crossing partitions towards them. When S comes as a set of nested partitions, a tree set, then its tangles will simply identify its locations (see Section 8.1), something one can readily do without tangles.

Finally, a word about the use of 'standard clustering algorithms' in our considerations above. Does this not beg the question, in that tangles are meant to offer just this, a new kind of clustering algorithm, but now we resort to existing such algorithms before we have even got started? In short, no. The standard clustering we use here need not be good, just simple and quick. It has little impact on the quality of our tangles, only on the time it takes to compute them: each use of standard clustering will result only in the choice of one partition of V to be included in S, the set providing the raw material for our tangles. The more partitions we adopt for S the better our tangles will be in the end: adding a 'bad' partition can at worst be unhelpful; it will most likely be weeded out immediately for having large order, in which case it will not affect the k-tangles obtained for small k.

Another way to view this use of standard clustering at the start is that our tangles emerge as an aggregate of many weak clustering inputs in one stronger clustering output. This is a standard technique in machine learning known as 'boosting' [28]. In our case those possibly weak clustering techniques are used, if at all, just to determine the partitions to which we then apply our tangles, not to replace those tangles.

Our discussion so far has centred on how to generate the first partition s of V that starts an iterative process of better and better partitions, partitions of decreasing order, until this process gets stuck. Let us now turn to this process itself: how exactly do we go about turning a given partition s into another partition s' of lower order?

The simplest way to do this is to try to move single elements of V across the partition, and check whether the order has gone down. With only a little more computational effort one can try to identify groups of elements on one side that form a cluster in some weak sense, and move the entire group across. In Section 10.2 we discussed an example of this, where those groups were the 'nodal domains' associated with a similarity function σ on V. The advantage of moving entire groups of points across the partition is that it may help us jump out of local minima that might exist with respect to moving single elements. For example, if a group of points is a cluster in the sense that they are pairwise similar but dissimilar to all the other points, then moving any one of these points across the partition will increase its order, but moving them all across may decrease it.

Another method for iteratively decreasing the order of a partition s was indicated towards the end of Section 10.1, cast in the framework of finding minimum cuts in graphs. This is a classical problem in network theory, and there are good algorithms available to find minimum cuts.

The sets S_k of partitions of V generated iteratively in this way may not be submodular in the sense of Section 7.5, a property assumed in the premise of all the theorems in Chapter 8. This would require S_k to contain, for every pair $r, s \in S_k$ of crossing partitions, also at least one corner from each of the two pairs of opposite corners. Algorithmically, it may well be possible to compute a tree of tangles, say, even if the corners required by submodularity do not always exist – but just happen to exist when the algorithm needs them (see Section 11.3).[3] Moreover, there are many tangle applications where all we are interested in are the k-tangles as such, rather than how they are structured by a tree of tangles.

But even in such cases it is good to beef up the set S we were given, or which we generated as outlined above, by adding for all $r, s \in S$ any sufficiently balanced corners $\vec{r} \wedge \vec{s} = A \cap B$ of specifications $\vec{r} = A$ and $\vec{s} = B$ whose order is no bigger than those of r and s, i.e., which satisfy

$$|\vec{r} \wedge \vec{s}| \leqslant \max\{|r|, |s|\}. \tag{11.1}$$

To see why it is generally a good idea to add low-order partitions to S in a partition scenario, recall our discussions in Section 7.4 of when a tangle 'dies', in the evolution of tangles, and does so not because it spawns new tangles when k increases but because k has reached a value where S_k contains so many partitions of V that it is impossible to orient them all as a tangle other than a boring principal tangle τ_v focussed on an element v of V. The same can happen now when we enlarge S_k not because k increases, but because we have generated another partition s of order less than k, either in our original generating procedure for S or retrospectively as a corner of partitions already in S. If the current S_k has a tangle τ that does not extend to a tangle of $S_k' = S_k \cup \{s\}$, then all this means is that τ was not a desirable tangle in the first place: it owed its tangle property, the fact that it contained no inconsistent triples, merely to the fact that it did not contain enough partitions to form such triples.

Put another way, the tangles that really matter in a partition scenario are the k-tangles, for k not too large, of the set S^* of *all* partitions of the ground set V.[4] We may not be able to compute such tangles, because S^* may be too large. But as soon as some k-tangle τ of a set $S \subseteq S^*$ fails to extend to a k-tangle of some set $S' = S \cup \{s\}$, we know that τ will not extend to a k-tangle of S^* either. In this case we call τ *fake*. It is a tangle for the wrong reasons: the intersections of 'all' its triples are large not because it approximates a k-tangle of S^*, one we would really care about, but because there are too few such triples in S to lend the tangle property of τ any substance. When the addition of s to S calls its bluff, we gladly discard τ.

11.2 Computing the tangles

Let us assume now that we have generated or are given a set S of partitions of V, have chosen an order function $S \to \mathbb{N}$ that assigns to every $s \in S$ a positive integer $|s|$, and have chosen a set $\mathcal{F}$ of subsets of $\vec{S}$ whose $\mathcal{F}$-tangles in $\vec{S}$ we wish to compute. Whenever we mention 'tangles' in what follows, they are meant to be $\mathcal{F}$-tangles for this $\mathcal{F}$. If we want to compute the most general tangles we let $\mathcal{F} = \mathcal{F}_1$; then all consistent specifications of whichever S_k we are targeting will be k-tangles of S.

The simplest approach to computing the tangles in $\vec{S}$ is to consider the elements $s_1, s_2, \ldots$ of S in increasing order $|s_1| \leqslant |s_2| \leqslant \ldots$, one at a time, and at step n compute all the tangles of $S^n := \{s_1, \ldots, s_n\}$. Having done this up to $n - 1$, say, we test at step n which of the two specifications of s_n can be added to which tangle of S^{n-1} to form a tangle of S^n. In other words, we check for both orientations $\vec{s}_n$ of s_n and every tangle τ of S^{n-1} whether $\vec{s}_n$ combines with elements of τ to form a set in $\mathcal{F}$: if it does, we cannot add $\vec{s_n}$ to τ. But if it does not, we can: then $\tau \cup \{\vec{s_n}\}$ is a tangle of S^n.

If neither $\vec{s_n}$ nor $\overleftarrow{s_n}$ can be added to τ in this way, we place $\tau \cap \vec{S_k}$ for $k = |s_n|$ on the list of tangles in $\vec{S}$ that our algorithm ultimately finds. Note that τ is a tangle of all of S_k, and thus a k-tangle of S, since $S_k \subseteq S^{n-1}$ by our choice of the enumeration $s_1, s_2, \ldots$ and $|s_n| = k$. In particular, if s_n is not the first element on our list of S that has order k, our tangle τ of S^{n-1} is discarded: while it induces a valid tangle of $S_k \subsetneq S^{n-1}$, which we keep, it is 'fake' as a partial tangle of S_{k+1}, because it does not even extend to a tangle of $S^n \subseteq S_{k+1}$.

As indicated, the above algorithm is only the most straightforward way to generate all the tangles in $\vec{S}$: it is possible to do this more efficiently. For example, if generating S uses similar amounts of computing power as building the tangles, we can do these in parallel, updating our partial tangles for each new s as it is generated. Or we can keep track of the pairs of elements of the partial tangles τ we have generated, the sets whose combination with a new features $\vec{s_n}$ must be tested for membership in $\mathcal{F}$, so that we do not have to compute them afresh every time we perform such a test. Another way to economize is to remember that, in order to store a partial tangle τ, it suffices to know its smallest elements (as subsets of V): if τ contains a feature $A \subseteq V$ and $A \subseteq A'$, it cannot contain $\overline{A'}$, since $A \cap \overline{A'} = \emptyset$ and so these two features are inconsistent. Hence if $\{A', \overline{A'}\}$ has small enough order that τ must orient this partition, it must contain A' rather than $\overline{A'}$. But rather than storing the information that $A' \in \tau$ (as well as $A \in \tau$), it is better to remember that $A' \supseteq A \in \tau$: this contains additional information, which may be useful in other contexts.[5]

Let us return to our observation from the end of Section 11.1 that adding low-order partitions to S can improve the quality of our tangles, and in particular help us weed out fake tangles. Is that also true when the partition added is a corner of two partitions already in S?

Consider the partitions $r = \{A, \overline{A}\}$ and $s = \{B, \overline{B}\}$ of V shown in Figure 11.1. Assume that their corner $\vec{t} := \vec{r} \wedge \vec{s} = A \cap B$ satisfies (11.1), e.g. that

$$|t|, |r| \leqslant |s| < k.$$

Can S have a k-tangle τ that fails to extend to a k-tangle of $S \cup \{t\}$?

One might think that this is unlikely. Indeed, as soon as τ contains $\overleftarrow{r}$ or $\overleftarrow{s}$, then adding $\vec{t}$ extends it to a tangle of $S \cup \{t\}$; recall that

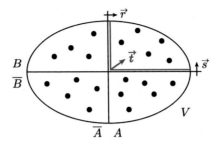

FIGURE 11.1.   Adding the corner $\vec{t} = \vec{r} \wedge \vec{s} = A \cap B$.

tangles are determined by their minimal elements, and in these cases $\overleftarrow{t}$ would not be minimal. But even if $\vec{r}, \vec{s} \in \tau$, it seems that adding t to S should be unproblematic for τ: while obviously $\tau \cup \{\overleftarrow{t}\}$ is not a tangle in this case, we should be able to extend τ by adding $\vec{t}$, the 'combination' of the features $\vec{r}$ and $\vec{s}$ that are already in τ. Indeed, if $\{\vec{r}, \vec{s}, \overleftarrow{t}\}$ is inconsistent then so was $\{\vec{r}, \vec{s}\} \subseteq \tau$ (a contradiction), since $\vec{r} \wedge \vec{s} \wedge \overleftarrow{t} = \vec{r} \wedge \vec{s}$.[6]

However, adding t to S can still reveal τ as fake. This is because $\vec{t}$ might form an inconsistent triple with two *other* elements $\vec{x}, \vec{y}$ of τ, such as the complements of two sets partitioning $A \cap B$. Then not only $\tau \cup \{\overleftarrow{t}\}$ but also $\tau \cup \{\vec{t}\}$ will fail to be a k-tangle of $S \cup \{t\}$, although τ was a tangle of S and $|t| < k$. Of course, replacing $\vec{t} = \vec{r} \wedge \vec{s}$ in $\{\vec{t}, \vec{x}, \vec{y}\}$ with $\vec{r}$ and $\vec{s}$ themselves will not change its inconsistency since, once more,

$$\vec{r} \wedge \vec{s} \wedge \vec{x} \wedge \vec{y} = \vec{t} \wedge \vec{x} \wedge \vec{y} = \emptyset\,;$$

but inconsistent quadruples are not forbidden in tangles. Thus, adding corners satisfying (11.1) to S may indeed help us weed out fake tangles. And as we noted already in Section 7.5, it will never create new tangles.

Moreover, in a setting where the partitions in S are interpretable, such as when S starts as a questionnaire (rather than a randomly generated set of partitions of V), and we wish to maintain this property when we enlarge S, then corners are the only new partitions we can add: these, too, have interpretations as Boolean expressions of the features in the original set $\vec{S}$, but arbitrary partitions of V do not.

Our above example was special in that the corner of r and s we added to S was not arbitrary. Rather, we picked a k-tangle τ of S, observed how it oriented r and s, and added to S the partition t underlying the infimum $\vec{t}$ of $\vec{r} = \tau(r)$ and $\vec{s} = \tau(s)$. If τ had any hope of extending to a tangle of $S \cup \{t\}$, it would have to be by $\vec{t}$ rather than $\overleftarrow{t}$, because $\{\vec{r}, \vec{s}, \overleftarrow{t}\}$ was inconsistent. If we add to S any of the other three corners of r and s, say t', then again only one specification of t' has a chance of extending τ to a tangle of $S \cup \{t'\}$: the specification $\vec{t'}$ that includes, as a subset of V, either $\vec{r}$ or $\vec{s}$ or both. However, adding t' to S (and $\vec{t'}$ to τ) will not increase or chances of exposing τ as fake: if $\vec{t'} \supseteq \vec{r}$, say, then for any inconsistent triple $\{\vec{t'}, \vec{x}, \vec{y}\}$ exposing $\tau \cup \{\vec{t'}\}$ as fake the triple $\{\vec{r}, \vec{x}, \vec{y}\}$ would also be inconsistent, so τ could not have been a tangle.

We therefore do not add all corners generated by partitions already in S, but only those that underly the infimum $\vec{r} \wedge \vec{s}$ of two features $\vec{r}, \vec{s}$ that jointly lie in some tangle τ of the current S. Some of the best ways to generate S, such as the spectral methods discussed in Section 10.2, return sets S of well-chosen partitions that are so small that we can often afford to add not only these corners, but also corners of corners etc. of this type, before we begin to compute the tangles in $\vec{S}$.

11.3 Computing a tree of tangles

Let $\vec{S}$ be a feature system of partitions of a set V. Assume that every partition s of V, not only those in S, has an order $|s| \in \mathbb{N}$.

Our aim is to compute a tree of the tangles in $\vec{S}$. Ideally, this would be a nested subset T of S that distinguishes all the tangles in $\vec{S}$ in the sense of Theorem 2: for every two distinguishable tangles in $\vec{S}$, say an ℓ-tangle ρ and a k-tangle τ with $\ell \leqslant k$, there would exist some $t \in T$ which ρ and τ specify differently, and this t would have minimum order among all such partitions in S. Clearly, any such t would have order less than ℓ, as otherwise ρ would not specify it.

If S consists of all the partitions of V and our order function $s \mapsto |s|$ is submodular, the existence of such a set T is ensured by Theorem 2. In practice, however, S will rarely contain all the partitions of V: it might be given as a questionnaire, or perhaps as a set of spectrally defined partitions of V, and in both these cases it will contain only a few of all the partitions of V.

The algorithm sketched below will work in such cases too. But before we describe it, we have to state more clearly what it should do: small sets S given as above might not contain any nested subsets of more than one element, let alone a nested subset that distinguishes all their tangles. Our aim, therefore, is to find T not inside S but to allow as elements of T also partitions of V that were obtained as corners of partitions in S, or as corners of corners, and so on. This does not mean that we allow arbitrary partitions of V in T; indeed, the set of partitions generated from S by taking corners will normally be much smaller than the set of all partitions.[7] Moreover, if S was a questionnaire, whose elements are questions and thus have meaningful interpretations, then corners of these are Boolean expressions of answered questions, which have meaningful interpretations too.

So we allow T to contain some partitions of V that are not in S. But what does it mean, then, that T distinguishes two tangles in $\vec{S}$?

Let S' denote any set of partitions of V obtained from S by adding corners, perhaps corners of corners etc., and assume that we wish to choose T from this set S'. Recall from Section 7.5 that every k-tangle of S extends to at most one k-tangle of S'; this is because at most one specification of a corner of two partitions specified by a tangle is consistent with that tangle. We can thus say that an element t of $T \subseteq S'$ *distinguishes* an ℓ-tangle ρ of S from a k-tangle τ of S if it distinguishes the unique ℓ-tangle $\rho' \supseteq \rho$ of $S \cup T$ from its unique k-tangle $\tau' \supseteq \tau$ and such tangles ρ' and τ' of $S \cup T$ exist. If they do not both exist, then ρ or τ was a fake tangle of S, one that does not extend to an ℓ-tangle or a k-tangle, respectively, of the set S^* of all the partitions of V, and we delete that fake tangle from the list of tangles in $\vec{S}$ we wish to distinguish by T.

Let us now turn to our algorithm for finding a tree of the tangles in $\vec{S}$, as in Theorem 2. Let us assume that we have already computed some set $\mathcal{T}$ of tangles in $\vec{S}$, those that we wish to distinguish by a tree set T. Let us assume also that these tangles are pairwise distinguishable, and that for every pair ρ, τ of these tangles we have found an element $s_{\rho,\tau}$ of S that distinguishes them, one which ρ and τ specify differently: say as $\overleftarrow{s}_{\rho,\tau}$ and $\overrightarrow{s}_{\rho,\tau}$, respectively. We further assume that $s_{\rho,\tau}$ distinguishes ρ from τ efficiently in S: that $|s_{\rho,\tau}| \leqslant |s|$ for every other $s \in S$ that distinguishes ρ from τ.

All this data is provided by our tangle-generating algorithm indicated in Section 11.2. Indeed, as we grow our set S of potential features in that algorithm, we only ever keep the maximal tangles we have found, discarding any earlier tangles that they induce. And for every pair ρ, τ of those maximal tangles we know an element of S that distinguishes them: the first s_n in our enumeration $s_1, s_2, \ldots$ of S that ρ and τ specify differently. In other words, while ρ and τ induce the same tangle on $s_1, \ldots, s_{n-1}$, they induce distinct tangles on $s_1, \ldots, s_n$, and we remember this first s_n that distinguishes them as $s_{\rho,\tau}$. Our notational convention is that $s_{\rho,\tau} = s_{\tau,\rho}$ is specified by ρ as $\overleftarrow{s}_{\rho,\tau} = \overrightarrow{s}_{\tau,\rho}$ and by τ as $\overrightarrow{s}_{\rho,\tau} = \overleftarrow{s}_{\tau,\rho}$. Since $s_1, s_2, \ldots$ were enumerated with increasing order, $s_{\rho,\tau}$ distinguishes ρ from τ efficiently in S. Let T initially denote the set of all these $s_{\rho,\tau}$.

This initial choice of T is already as small as one can hope to make any set that distinguishes all the tangles in $\mathcal{T}$: it has size at most $|\mathcal{T}| - 1$. However, the partitions in this initial choice of T will typically cross. Our aim that T shall be a tree set, however, is crucial to a number of tangle applications. One consequence is that its tangles extend (uniquely) to tangles of S. Another is that it enables us to associate elements of V with tangles of T, which turns these tangles into traditional clusters in V; see Sections 14.1 and 14.2. And both these properties together form the basis for tangle-based predictions as discussed in Section 12.1.

The task of our algorithm, therefore, is to 'uncross' any partitions in T that are not nested: to turn T into a tree set by replacing, at each of a sequence of steps, one of a crossing pair of elements of T with a corner of the two. The aim is that, after every such replacement, the amended set T contains fewer crossing pairs of partitions but still distinguishes all the tangles in $\mathcal{T}$ efficiently. This efficiency will, in fact, become stronger, as it will always refer to our growing current set of partitions: initially to S, but as we add new partitions (corners) to build our tree set, these too compete for efficiency as distinguishers of any pair of tangles in $\mathcal{T}$.

Let us look at a single step of this algorithm in more detail. Let $r = s_{\rho,\tau}$ and $s = s_{\varphi,\psi}$ be crossing elements of our current set T which efficiently distinguish tangles $\rho, \tau \in \mathcal{T}$ and $\varphi, \psi \in \mathcal{T}$, respectively. Let us assume that $\ell := |r| \leqslant |s| =: k$. The algorithm now tries to find a corner of r and s that either distinguishes ρ from τ and has order at most ℓ, or which distinguishes φ from ψ and has order at most k; we then replace r or s in T with this corner (Figure 11.2).[8]

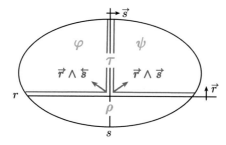

FIGURE 11.2.   The partition s distinguishes φ from ψ, but r does not.

When the algorithm adds a corner t to T, it also adds t to the set S of partitions whose tangles are computed, though without deleting any other partition from this set. It then updates, for every $m > |t|$, every m-tangle θ that has so far been computed for the current version of S, by adding to it any orientation of t that does not form an inconsistent triple with θ and thus extends θ to a tangle of the new S. If $m > k$, this will happen for at most one orientation of t, since t is a corner of two partitions already oriented by θ (see Section 7.5). If m is smaller, it can happen that both $\vec{t}$ and $\overleftarrow{t}$ extend θ to an m-tangle of the new S. We then replace θ with these two extensions θ' and θ'' in our collection of tangles. In either case, the algorithm also updates to any extensions of θ the distinguishers between θ and other tangles of S, by replacing them with t if t distinguishes them too but has smaller order. If neither orientation of t can be added to θ then θ was fake. Then the algorithm deletes θ from $\mathcal{T}$, and its distinguishers from T.

If the algorithm cannot find such a corner t of r and s, but the order function it uses is submodular, then at least one of the four tangles $\rho, \tau, \varphi, \psi$ was fake; we shall see why below. It deletes these fake tangles from $\mathcal{T}$, and their distinguishers from T.

The algorithm ends when T is nested.

There are two claims in our description of the above algorithm that need justification. One is that, if the algorithm cannot find a corner as envisaged, then one of the tangles considered was fake. To see this, let us look at Figure 11.2. If one of the two upper corners, t say, has order at most $k = |s|$, it will distinguish unique extensions of φ and ψ to $S \cup \{t\}$ and can thus replace s as an efficient φ, ψ-distinguisher. If both upper corners have order greater than k then, by the submodularity of our order function, both lower corners have order less than $\ell = |r|$. One of these, or both, can be added to S and replace r as an efficient ρ, τ-distinguisher, as long as we can orient that corner without creating an inconsistent triple with other elements of ρ. But if t is one of these corners and both $\rho \cup \{\vec{t}\}$ and $\rho \cup \{\overleftarrow{t}\}$ are inconsistent, then ρ was fake: it does not extend to a $|\rho|$-tangle of all the partitions of V, where $|\rho|$ is the order of ρ, because it does not even extend to a $|\rho|$-tangle of $S \cup \{t\}$.

The other (implicit) claim that needs justification is that our algorithm terminates: that it does not run forever in cycles, replacing elements of T with other partitions which in turn are replaced by the original ones later. A simple way to ensure this is to change the orders of our partitions just very slightly so that no two of them have exactly the same order.[9] Since the order of any corner t that replaces a partition s in T satisfies $|t| \leqslant |s|$, this inequality will then be strict, which ensures that our process ends.

A more sophisticated algorithm than the above has been proposed by Grohe and Schweitzer [24]. It turns the proof of Theorem 2 given in [13] into an algorithm with a polynomial running time; however, it is not intended for practical applications.[10] Our algorithm is a simplification of an algorithm due to Elbracht, Kneip and Teegen [18]. It has an exponential worst-case running time but is quite fast in practice [1].

11.4 Computing a tree set certifying that no tangles exist

Let S be a set of partitions of a set V, and let $\mathcal{F} = \mathcal{F}_n$ for some $n \geqslant 1$. Our set S might be the set S_k^* of all partitions of V of order less than some given k, or it might come straight from a questionnaire answered by the people in V, or a combination of the two.

Theorem 3 in Chapter 8 tells us that, if $\vec{S}$ is submodular, then S either has an $\mathcal{F}$-tangle or contains a tree set over $\mathcal{F}$. However we may well be interested in finding one of these two structures also if $\vec{S}$ is not submodular: then it is not guaranteed that we find one or the other, but it might still be the case. In what follows we shall not assume that $\vec{S}$ is submodular.

We saw after Theorem 3 that the two alternatives it offers cannot happen at once: as soon as we have found a tree set over $\mathcal{F}$ in S we know that S has no $\mathcal{F}$-tangle. Our algorithmic task now is the converse: assuming that our tangle search from Section 11.2 has failed to produce an $\mathcal{F}$-tangle of S, we now wish to find a tree set over $\mathcal{F}$ in S as a certificate that indeed no tangle exists.

Very informally, such an algorithm might do the following. It builds a list L of features that any $\mathcal{F}$-tangle of S must contain, given the triples in $\mathcal{F}$ which it must *not* contain. If, at some point, this list L gets 'overdetermined' in that it contains both specifications $\vec{s}$, $\overleftarrow{s}$ of some $s \in S$, we know that S has no $\mathcal{F}$-tangle. Our list L will be supported by bookkeeping from when we built it, which will then enable us to piece together some of the features in L to form a tree set over $\mathcal{F}$.

To get started, our algorithm looks for singleton sets $\{\overleftarrow{s}\} \in \mathcal{F}$, those with $\overleftarrow{s} = A \subseteq V$ with $|A| < n$, and puts their inverses $\vec{s}$ in L. From now on it repeats the following simple step. It looks for stars $\sigma \in \mathcal{F}$ of which all but one element, $\overleftarrow{s}$ say, are already in L and hence lie in any

$\mathcal{F}$-tangle of S, but $\vec{s} \notin L$. This forces $\vec{s}$, too, to lie in any $\mathcal{F}$-tangle of S, and so the algorithm adds it to L. It also remembers σ as the *reason* for $\vec{s}$ being in L. It does this, in the current step, for all such σ and $\overleftarrow{s}$. This process terminates, because it adds a new feature to L at every step.

If our algorithm, at some point, places both specifications $\vec{r}$ and $\overleftarrow{r}$ of some $r \in S$ in L, it stops and builds a tree set over $\mathcal{F}$, as follows. Starting with $\vec{s} := \vec{r}$, we build an oriented tree set $\vec{T}_{\vec{r}} \subseteq L$ with $\vec{r}$ as its smallest element. Having added a feature $\vec{s}$ to $\vec{T}_{\vec{r}}$ we look up the reason $\sigma \in \mathcal{F}$ of why $\vec{s}$ was put in L. We then put all the elements of $\sigma \setminus \{\overleftarrow{s}\}$ in $\vec{T}_{\vec{r}}$, remember the star σ as a location of T, and repeat this step separately with all the newly added elements of L as $\vec{s}$. A branch of this process ends when the reason for why we put a feature $\vec{s}$ in L is the singleton star $\sigma = \{\overleftarrow{s}\} \in \mathcal{F}$, in which case the set $\sigma \setminus \{\overleftarrow{s}\}$ to be added to $\vec{T}_{\vec{r}}$ is empty. When the construction of $\vec{T}_{\vec{r}}$ is complete, because all the branches of this process have ended, we run a similar process starting with $\vec{s} = \overleftarrow{r}$, to form an oriented tree set $\vec{T}_{\overleftarrow{r}}$ whose smallest element is $\overleftarrow{r}$. Our overall tree set $T \subseteq S$ over $\mathcal{F}$, then, is the set of potential features underlying the set $\vec{T}_{\vec{r}} \cup \vec{T}_{\overleftarrow{r}}$ of features. Its locations are the stars $\sigma \in \mathcal{F}$ that provided our reasons for putting features from $\vec{T}$ in L.

It is not hard to prove that if S contains a tree set over $\mathcal{F}$ then our algorithm will find one, either this or another, by running until it stops for the reason indicated above: that it has placed a pair of inverse features in L.[11] If $\vec{S}$ is submodular, Theorem 3 ensures that this happens whenever S has no $\mathcal{F}$-tangle.

If $\vec{S}$ is not submodular, it can happen that S has no $\mathcal{F}$-tangle and yet contains no tree set over $\mathcal{F}$.[12] In such cases, however, we can still run a modified version of our algorithm that can – but does not always – produce certificates for the non-existence of an $\mathcal{F}$-tangle of S that are as useful as those provided by Theorem 3.

To see what these are, we need the concept of a graph-theoretical tree. We say that such a tree is an *S-tree over $\mathcal{F}$* if its nodes are elements of $\mathcal{F}$, not necessarily stars, its edges are elements of S such that any edge joining two nodes $F, F' \in \mathcal{F}$ has exactly one specification in F and the other in F', and conversely every element $\vec{s}$ of a node F is a specification of an edge at F.

The edges of an S-tree over $\mathcal{F}$, all elements of S, need not form a tree set; indeed, even the elements of a single set $F \in \mathcal{F}$ can cross. However, an S-tree T over $\mathcal{F}$ is still a certificate that S has no $\mathcal{F}$-tangle.

To see this, we just have to adapt the easy proof of the corresponding fact in Theorem 3 given there, as follows. Suppose S has an $\mathcal{F}$-tangle τ. The orientation $\tau(s)$ of every edge s of T can be made visible by placing an arrow on this edge $s = FF'$: towards F if $\tau(s) \in F$, and towards F' if $\tau(s) \in F'$. Then we follow these arrows along the edges of T until we get stuck at a 'sink', a node F that contains $\tau(s)$ for every edge s at F.[13] Then $\tau \supseteq F \in \mathcal{F}$, which contradicts our assumption that τ is an $\mathcal{F}$-tangle. Thus, any S-tree over $\mathcal{F}$ is a certificate for the non-existence of any $\mathcal{F}$-tangle of S.

Moreover, if there exists an S-tree over $\mathcal{F}$ then our earlier algorithm, with just a tiny adjustment, will find one. The adjustment is that all sets $F \in \mathcal{F}$ are now eligible as 'reasons' for adding a feature $\vec{s}$ to L, not just those that are stars. With this adjustment, everything works as before. Just note that if $\vec{S}$ is not submodular then there need not, in general, exist either an $\mathcal{F}$-tangle of S or an S-tree over $\mathcal{F}$, even for $\mathcal{F}$ of the form $\mathcal{F} = \mathcal{F}_n$ as considered here.[14]

If $\vec{S}$ is submodular, on the other hand, the alert reader may have become suspicious. We have seen that the existence of an S-tree over $\mathcal{F}$, even one whose edges include many crossing elements of S, implies that S has no $\mathcal{F}$-tangle. If $\vec{S}$ is submodular, this implies by Theorem 3 that S has a tree set over $\mathcal{F}$. But that is the same as an S-tree over $\mathcal{F}$ whose edges do *not* cross – which in turn is the same as an S-tree over the set $\mathcal{F}^*$ of just the stars in $\mathcal{F}$. So it seems that, for submodular $\vec{S}$, the existence of an S-tree over $\mathcal{F}$ implies the existence of an S-tree over just $\mathcal{F}^*$.

Can this be true? After all, there does not appear to be any obvious way to turn an arbitrary S-tree over $\mathcal{F}$ into one over $\mathcal{F}^*$.

It is indeed true. To see why, let us consider the corresponding reverse implication on the tangle side of the equivalence in Theorem 3: let us show that any $\mathcal{F}^*$-tangle of S is in fact an $\mathcal{F}$-tangle.

Consider any $F \in \mathcal{F}$. Its elements may cross, but we can turn F into a star σ whose interior, the intersection of all its elements viewed as subsets of V, is the same as the corresponding intersection for F: a star σ such that $\bigcap \sigma = \bigcap F$. Moreover, we can find such a star σ with $|\sigma| \leqslant |F|$.

To see this, consider a pair $\vec{r}$, $\vec{s}$ of crossing elements of F. Since $\vec{S}$ is submodular, it also contains one of the corners $\overleftarrow{r} \wedge \vec{s}$ and $\overleftarrow{s} \wedge \vec{r}$. If it contains $\overleftarrow{r} \wedge \vec{s}$, we replace $\vec{r}$ in F with the inverse $\vec{r} \vee \overleftarrow{s}$ of this corner. Note that this replacement does not increase $|F|$ or change $\bigcap F$: although the new set $\vec{r} \vee \overleftarrow{s}$ is a superset of the set $\vec{r}$ it replaces, the additional subset $(\vec{r} \vee \overleftarrow{s}) \smallsetminus \vec{r}$ lies in $\overleftarrow{s}$, and is therefore deleted again when the intersection $\bigcap F$ is formed with $\vec{s} \in F$. Similarly, if $\vec{S}$ contains $\overleftarrow{s} \wedge \vec{r}$, we can replace $\vec{s}$ in F with the inverse $\vec{s} \vee \overleftarrow{r}$ of that corner. As before, this does not change $\bigcap F$ or increase $|F|$. We apply one of these changes, not both. It is not hard to show that repeating this uncrossing step some finite number of times turns F into the desired star σ.[15]

Moreover, since $\vec{r} \vee \overleftarrow{s} \supseteq \vec{r}$ and $\vec{s} \vee \overleftarrow{r} \supseteq \vec{s}$, and since any tangle of S also contains all the supersets in $\vec{S}$ of its elements (as containing their inverse would make it inconsistent), any $\mathcal{F}^*$-tangle τ of S that includes F will also include the star σ we obtained from F by uncrossing its elements. But since $\mathcal{F} = \mathcal{F}_n$ and $\bigcap \sigma = \bigcap F$, this star σ lies in $\mathcal{F}^*$. This contradicts our assumption that τ is an $\mathcal{F}^*$-tangle. Thus, τ does not include our arbitrary $F \in \mathcal{F}$, and hence is an $\mathcal{F}$-tangle.

This connection between $\mathcal{F}$-tangles and $\mathcal{F}^*$-tangles somewhat demystifies our earlier observation that, somehow, the existence of an S-tree over $\mathcal{F}$ must imply the existence of an S-tree over $\mathcal{F}^*$, although we could not see how to construct the latter from an instance of the former. If there exists an S-tree over $\mathcal{F}$ then S has no $\mathcal{F}$-tangle; then, by uncrossing, it does not even have an $\mathcal{F}^*$-tangle; then by Theorem 3 it contains a tree set over $\mathcal{F}^*$ which, in essence, is an S-tree over $\mathcal{F}^*$.

Note that the above argument relied heavily on our assumption that $\mathcal{F}$ has the form of $\mathcal{F} = \mathcal{F}_n$. A more general version of the tangle–tree duality theorem, due to von Bergen [2], works for other $\mathcal{F}$ too and been implemented in the software package [1].

Part IV

Applying Tangles

Back to the examples

In this Part we return to our collection of examples from Part II. We shall review each of those examples from our new perspective of the deeper understanding we gained in Part III of what tangles are.

While our discussions in Part II centred on how the notion of tangles as such plays out in each context, they will now also include parameters that influence this. This begins with how to choose or generate the set S of partitions, or potential features, of our objects in V. Another topic is how to choose, explicitly or generically, an order function on S that is

suitable in the given context. This has implications for the corresponding evolution of tangles introduced in Section 7.4, and their hierarchy
displayed by Theorem 2. In some cases we shall note that, perhaps
surprisingly, the feature system dual to the one that comes to mind first
can illuminate the tangle application at hand.

Not all these considerations will be made explicit in each of the contexts of tangle applications that we revisit. The exposition in Part IV
assumes familiarity with Part III throughout, and it includes cross-refer-
ences only when these are particularly relevant. This is because the examples we discuss are still for motivation and illustration purposes only:
they are not intended as blueprints for concrete tangle applications, even
in the contexts discussed, that can be performed out of the box. The aim
is, rather, to indicate a multitude of possible applications by a colourful
collection of synthetic examples that might, perhaps, be used as templates, but which primarily serve to illustrate a range of aspects that
are likely to come up in any concrete application of tangles designed by
experts in the relevant field.

We do, however, discuss in most cases what our theorems from
Chapter 8 imply for the given context. In particular, we shall look at
the role played in that context by the tree sets of partitions offered by the
tree-of-tangles theorem or the tangle–tree duality theorem. All 'tangles'
mentioned will be $\mathcal{F}$-tangles: typically, but not necessarily, with $\mathcal{F} = \mathcal{F}_n$
for some integer n. Whenever this is left unspecified they are $\mathcal{F}_1$-tangles,
the most general kind.

12

Applying tangles in the natural sciences

In this chapter we revisit the three examples of tangle applications in the sciences that we discussed in Chapter 4: building an expert system to assist doctors in the process of medical diagnosis; identifying new viruses or proteins, and their phylogenetic trees, from their sequences of nucleotides or amino acids in a large pool of molecules; and developing new drugs that target several pathogens at once. As mentioned before, these are just toy examples intended to illustrate a diversity of potential uses of tangles, which will have to be developed by, or in collaboration with, experts in the respective field of application.

To start this chapter, we pick up a topic that transcends our division of application fields between the sciences: the topic of predictions. This was treated informally in Section 3.3, in a unified way for all fields of potential tangle applications. Following our more formal treatment of tangles, in particular the possibility to have trees of tangles as provided by Theorems 1 and 2, we can be a little more specific now.

12.1 Predictions

This section could be equally well placed in the next chapter: predictions are as relevant in the social sciences as in the natural sciences. In fact, to make our discussion more intuitive we shall adopt the language of the questionnaire example from Section 1.2, rather than the terminology closer to the natural sciences as used in Section 1.1. Thus, S is a questionnaire (rather than a set of possible measurements), and V is a set of people that have answered these questions (rather than a set of phenomena, each with a known set of readings for the measurements in S).

We are now interested in how an unknown person might answer the questions in S. This could be formalized by assuming that V is a sample of some larger population P of people whose views on S interest us. We only know the answers given by the people in V, but have reason to believe that these represent the people in P well, and are now interested in the likely views on S of a person in $P \smallsetminus V$.

Although we shall need to compute the tangles of S in order to make our predictions, we will not need to 'understand' the tangles we have computed, to 'make sense' of them: the tangle-based process of predicting the answers to S of our unknown person from P can be carried out entirely mechanically on the basis of the data available to us, the way in which the people in V have answered S.

As envisaged in Section 3.3, our aim is to use a tangle-distinguishing set T as discussed in Section 8.2 to make such predictions. For simplicity, let us do this for the scenario of Theorem 1. We assume that we have computed the tangles of S for some $\mathcal{F}$ of our choice, and that we have also computed our tangle-distinguishing tree set $T \subseteq S$.

The idea now is that, since tangles express views on S that are typical for the people in V, any tangle should be a better guess for the overall views on S of our unknown person $p \in P \smallsetminus V$ than a random specification of S. If, in addition, we already know this person's views on T, then these would even tell us which tangle of S to choose as our guess: the unique tangle of S that induces this specification $p(T)$ of T, as in Theorem 1.[1]

When we test this hypothesis on data from an existing survey S carried out on a population V, there are a few simple benchmarks for our tangle-based predictions to beat.

If we know nothing at all about the people in $P \smallsetminus V$, two reasonable options for guessing their views based on those of the people in V – let us refer to this guess as the specification σ of S – are as follows. One is to choose the entire specification σ as the one that has greatest support in V, the one for which $|\{\, v \in V : v(S) = \sigma \,\}|$ is largest. However, this cuts things a bit fine: it depends only on how many $v \in V$ chose *exactly* this σ as their $v(S)$, no matter how popular σ is 'by and large'. A hugely popular tangle τ of S, for example, say one such that every $v \in V$ agrees with τ on 90% of the questions, would not be chosen as σ, simply because no $v(S)$ agrees with τ entirely.[2]

At the other extreme we could take as $\sigma(s)$, separately for each $s \in S$, the majority opinion of the people in V, letting $\sigma(s) = \vec{s}$ if more $v \in V$ chose $\vec{s}$ as $v(s)$ than $\overleftarrow{s}$. While this is clearly the best possible guess (from the data we have) for each individual s, it may be less good as an overall guess at an entire specification of S. Indeed, the collection of the most popular views on single questions s may, in its entirety, even be inconsistent.

For example, if $S = \{s_1, s_2, s_3\}$ with the $\vec{s_i}$ forming a tripartition of V into equally-sized parts, then the majority vote of all $v \in V$ on each s_i will be $\overleftarrow{s_i}$, where two thirds of V lie. The overall guess σ consisting of the three $\overleftarrow{s_i}$, however, would not agree with the views of any $v \in V$ entirely. Indeed, it would disagree with every $v \in V$ on exactly one s_i, that for which $v \in \vec{s_i}$, while agreeing on two out of three questions with all the other v. By comparison, the three tangles containing $\vec{s_i}$ for one i and $\overleftarrow{s_i}$ for the other two each agree entirely with a third of the people in V, but agree with the rest on only one out of three questions. Out of context it is impossible to say which is better, but there will likely be applications where the tangle-based prediction wins.

If our unknown person p has answered the questions in T, a natural benchmark is to pick an arbitrary set $S' \subseteq S$ of size $|S'| = |T|$, solicit from p a set σ' of answers on S', and then to fashion σ after the views on S of those $v \in V$ that agree with p on S', those with $v(s) = \sigma'(s)$ for all $s \in S'$. This could be done, for example, in one of the two ways discussed above, but now based only on the views of this select subset of people in V. Our hypothesis now is that choosing a tangle of S that extends $p(T)$ is a better guess for σ than one that extends $p(S')$.

What we have discussed so far are possible ways of testing the idea of tangle-based predictions against real-world data, much as one would test a theory in the natural sciences. But if this is mathematics, can we not *prove* that tangle-based predictions are good, at least better than making arbitrary guesses?

The short but definitive answer is 'no': no amount of tangle theory, or indeed of any mathematics, can prove this. Apart from the fact that we would have to agree on what 'better' should mean, this is simply because we have not made any assumptions about how the people in V answered the questions in S.

Such assumptions can be made in the form of a probabilistic model. For example, we might assume that the views $v(S)$ on S held by the people $v \in V$ satisfy a certain probability distribution. Once we have made such an assumption, we can 'forget V' and, for the purpose of our prediction, work directly with the abstract probability space $\{yes, no\}^S$ of all the specifications of S. This will allow us to re-interpret S as a set of 'potential features' also of the elements of P, even if these potential features are not known individually as partitions of P, and then prove that tangle-based predictions are better or worse for this distribution than other predictions.

Such models of how the views $v(S)$ on S of the people in V are distributed cannot, however, be established once and for all: they will have to be based on intimate knowledge of the intended application and can only be devised by, or in collaboration with, experts in those fields.

So let us leave this discussion here, and just remember that modelling tangle predictions in a rigorous way remains a challenging interdisciplinary problem. A problem, however, which we do not have to solve in order to make such predictions: all we have to do is compute the tangles of S and its tangle-distinguishing subset T.

12.2 Expert systems: fuzziness versus structure in medical diagnostics

In Chapter 4 we discussed tangles as an approach to building expert systems for the process of medical diagnosis. While this is just one of numerous potential examples, see the end of Section 4.1, it is both particularly important and particularly fitting, given how imprecise data about humans can be. Let us continue with this example here.

The basic idea is that illnesses, known or unknown, should show up as tangles in any large enough set of diagnostic measurements: informally, an illness would be understood as a 'typical' collection of readings of a set S of measurements that can be performed on patients, and thus be a tangle of S by the informal definition of 'tangle' from Section 2.3.

To set up tangles more formally we need a set V of 'points', so that S becomes a set of potential features of these points, a set of partitions of V. In our case, V will be the set of people whose medical conditions we plan to use to determine whether or not a set of readings of the measurements in S is typical, and therefore constitutes a tangle. Ideally, this would consist of all humans on medical record, past and present. More practically, it might be a large set of individuals whose medical data are known to whoever undertakes to build this tangle-based expert system to assist with medical diagnosis.

The idea now is to compute the tangles of S, or more broadly the various tangles in $\vec{S}$ for a suitable order function on S, from the values of $v(s)$ for all $s \in S$ and all $v \in V$, which we assume to be in our database, and then to compute the tree of tangles $T \subseteq S$ as in Theorem 1 or 2. The collection of all these tangles, and of the locations of T at which they live, will then form the foundation for our expert system. With each tangle we might also store some potentially interesting parameters, such as its order k and its agreement value n.

Note that we have to compute these tangles only once: when we build our expert system. Those that later come to use it will no longer have to compute any tangles. However, we might allow for the system to be updated as more data becomes available. Then the server that runs the expert system will have to re-compute those tangles occasionally, as more data comes in. But let us assume for now that we build the system only once, and then apply it many times. This would happen as follows.

Let p be a patient whose illness we seek to determine. Let us assume that no medical data on p is known. Our expert system should tell us which diagnostic measurements we have to perform on p, and in which order. The aim is to identify a tangle τ of S, of order as high as possible, that is close to $p(S)$ in the sense of Section 6.1.

So far, this is quite similar to what we did in Section 12.1. The difference is that for the 'unknown person' p there we had to guess how they would answer the questions in $S \smallsetminus T$, whereas now we can test any such guess by examining our patient p. The reason we still need tangles is that it would be impractical to test p on the entire set S. Tangles can help us reduce the number of tests needed: to $p(T)$ first, plus $p(s)$ for any s needed to confirm our guess of τ made on the basis of $p(T)$.

Traditionally, the first 'measurements' performed on any patient consist of a record of their medical history and a list of the conditions that made them seek medical help. These will be included in S. Which subset of S they are, let us call it S', may depend on our patient p. We assume that we know how p specifies this S'.

Next, our expert system will ask us to test p on T. In fact, it will ask us to test p on only those t of T whose specification is not already determined by earlier measurements, given our assumption that $p(S)$, and in particular $p(T)$, is consistent.[3] Once we know $p(s)$ for some $s \in S'$ or $s \in T$, we only have to measure $p(t)$ for any $t \in T$ such that both $\vec{t}$ and $\overleftarrow{t}$ are consistent with $p(s)$, i.e., have non-empty intersection with it as subsets of V.

If the expert system suggests the measurements in T in the right order, this shortcut can be substantial, so that we have to perform far fewer than $|T|$ measurements to determine $p(T)$. Elements of T whose specification determines that of many other elements of T by consistency are those that are 'central' in T: elements t such that both $\vec{t}$ and $\overleftarrow{t}$ have many subsets $\vec{t'}$ for other $t' \in T$. We let our system ask those t first.

If T was chosen minimal, which we may assume, then $p(T)$ has the form $\tau \cap \vec{T}$ for some tangle τ of S, or for some maximal tangle τ in $\vec{S}$ in the setup of Theorem 2. This tangle τ encodes the illness with which we provisionally diagnose our patient p.

The correctness of this diagnosis τ hinges on the correctness of $p(t)$ for every single measurement $t \in T$, much like the traditional clinical method of diagnosis 'by exclusion': changing a single $\vec{t} \in \tau \cap \vec{T}$ to $\overleftarrow{t}$ will change τ to a different tangle of S. The reason our tangle approach is still more robust than the exclusion method is twofold.

First, T is much smaller than S. A measuring error on some $t \in T$, therefore, is less likely than one on some $s \in S$ tested traditionally, at least if we assume that those s are less carefully chosen than our t and are therefore more numerous before a definitive diagnosis emerges.

Second, if an erroneous $p(s)$ occurs in the traditional exclusion process, the result will often be an inconsistent specification $p(S'')$ of the subset set S'' of measurements performed. In that case the diagnostic process simply fails: 'the doctor couldn't find anything wrong with me' is nothing but a re-wording of the statement that no known illness specifies S'' as $p(S'')$.

In our case, a measuring error may also yield an inconsistent overall result. But the inconsistency will happen within the small set T of measurements. We can then re-take these measurements until $p(T)$ is consistent. Once we have established $p(T)$, there will be a unique tangle τ of S, or in $\vec{S}$, such that $p(T) = \tau \cap \vec{T}$. If this diagnosis τ looks at all suspicious, or as a precautionary measure, we can go on to test p on more $s \in S$, specifically for the correctness of τ. Since τ specifies all of S in our database, we know for every such s whether the value of $p(s)$ returned is compatible with our diagnosis τ.

Let us now address two more specific questions that come up when we first design our expert system: how to choose our order function, if any, and how to set the agreement parameter for our tangles τ of S: the minimum number n of 'points' $v \in V$ we require for every three features $\vec{s}$ of τ to satisfy $v(s) = \vec{s}$.

Increasing the agreement parameter n makes it harder for a specification of S to count as a tangle. The algorithms designed to detect tangles in a dataset often allow the user to choose n interactively, watching the number of tangles detected decrease as n is increased. The user can then decide when to stop, so as to get the most reasonable number of tangles for their work.

In our present application, however, the aim is not to detect tangles, but to fit tangles already determined to a new patient p and their $p(S)$. When we determine those tangles at the time we build the expert system, we should not fix any value of n but allow n to vary, yielding different sets of tangles. But we can save, for every tangle we compute, the maximum n for which it is an $\mathcal{F}_n$-tangle. Tangles with large agreement n simply encode illnesses that are more common, and are therefore more likely to be the correct diagnosis for our patient p than tangles of low agreement.

Note that n typically varies with the order k of the tangles we consider: as k increases, their agreement value n will decrease, simply because S_k gets larger. Since every k-tangle induces ℓ-tangles for all $\ell < k$, those k-tangles will typically encode more specific illnesses, which in turn will be less common. For example, the common cold may correspond to an ℓ-tangle ρ for some relatively small value of ℓ, and this tangle will have large agreement n, since many people have a common cold. But τ may split into various k-tangles for $k > \ell$, perhaps tangles for particular strands of virus each triggering symptoms of a a common cold. For each strand, there will be fewer people in V that had this type of cold, and so its tangle τ will have smaller agreement value than ρ.

But how should we choose our order function on S?

There are a number of different criteria for making this choice, which depend on the angle from which we view this task. From a mathematical point of view, we will get the clearest tangles from generic order functions as discussed in Chapter 9. These will generally assign low order to measurements that mark clear bottlenecks in V: measurements for which most illnesses have some common reading on most of the people with that illness. By contrast, measurements that yield unspecific readings for many illnesses will split the sets of people with those illnesses evenly, and thus receive high order in order functions such as (O1).

This is generally a good thing: if we assign low order – less than k, say – to too many partitions that evenly divide a set X of people in V that have a particular illness, then no k-tangle will correspond to that illness (see Sections 2.4 and 7.3). We should be aware of this if we choose to define our order function manually. However there are reasons to do this, such as the following considerations.

From a practical point of view we would like the tests we perform on our patient p to be as simple as possible. So we prefer trees of tangles T whose measurements $t \in T$ are easy to perform: taking a patient's pulse, blood pressure, weight and so on. If we assign those tests low order, they will be more likely to appear as tangle-distinguishers in T.

But we also could, and should, ask medical experts which measurements we should regard as more fundamental than others, in the sense that they are likely to distinguish many different illnesses. For example, body temperature is a measurement that is traditionally taken early on in any diagnostic process. And there are good reasons to do so also from the perspective of tangles. It can help us distinguish an infection from cancer, say, or a viral from a bacterial infection. But it is less likely to distinguish between different forms of cancer or different types of bacterial infection. So body temperature does what we look for in a good basic distinguisher that should receive low order: one that splits V fairly evenly but is unlikely to cut right through any set of people that have the same illness. It would probably be assigned low order by our generic order functions from Chapter 9 without the need of expert interference.

By contrast, a rash typical for scarlet fever will also be given low order by a generic order function such as (O1), because the condition is rare and distinguishes only the few patients that do have scarlet fever from the others. But we might not want to assign it low order, because such a rare measurement – even if easy to perform, but worse if not – would clutter the advice given to doctors by our system in the early stages of diagnosis.

However, even low-order measurements will be suggested early by our expert system only if they are relatively central elements of T, see earlier. But measurements that distinguish only one illness from all others, such as a scarlet rash, cannot be central in T: they do not distinguish enough tangles.

In summary, choosing a suitable order function is an important task that will be part of the process of building of our expert system. It is one of the key channels through which medical expertise can, and should, inform this process. At the same time it will require expertise on tangles, to ensure that it gives rise to meaningful distinguishable tangles that correspond to illnesses.

12.3 DNA and protein tangles: aligned and alignment-free

In Chapter 4, Section 4.2, we applied tangles to look for sequences of letters A, T, G and C, as they define DNA nucleotides, that were 'typical' for a sample of DNA molecules we sought to identify, classify, or relate to each other. As pointed out at the time, the same approach can be applied to sequences of amino acids in proteins, and this continues to be the case now.However we shall continue with the DNA example here, to make it easier to go back and forth between the corresponding sections.

Our tangle framework in Section 4.2 was the questionnaire setting: every relevant position in a DNA molecule corresponded to four yes/no 'questions', namely, whether in that position it featured A, T, G or C. A tangle of the set S of these positions would be a particular specification of S, a typical – if hypothetical – concrete sequence of bases, one for each position.[4] We shall begin this section by working out this approach in more detail, discussing order functions, the corresponding hierarchy of tangles in $\vec{S}$ (rather than tangles of the entire set S), and see how the tree of tangles from Theorem 2 can be interpreted as a phylogenetic tree of the organisms in our sample. We shall also discuss what Theorem 3 can tell us about our sample when there are no tangles of the kind we were looking for.

Since it is crucial in this setup that we have well-defined 'positions' in our DNA molecules, we must assume for this approach that the DNA molecules in our sample are aligned. Depending on the context of our application, this may not always be possible.[5] At the end of this section we discuss how tangles can be used for the same aim without alignment.

Let us start, then, with our questionnaire-type setting, in which every $s \in S$ is a pair (p, B) of a relevant position p in our DNA and a base B that may or may not be present at position p. For simplicity, let us look for order functions on S where the order of $s = (p, B)$ depends only on p, not on the choice of $B \in \{A, T, G, C\}$. Given s, we may then refer to this position p as $p(s)$.

As always, the idea is to assign low order to those $s \in S$ whose $p(s)$ is particularly fundamental, or important for the aims of our investigation. These two may not be the same thing: a position would be fundamental if it allowed us to distinguish between species that have no recent common ancestor, but it may be important also if it indicates a rare mutation that is known to be responsible for some particular illness. All we can say here is that, for this approach, the order function on S has to be hand-designed by the experts.

Recall that, as $S_1 \subseteq S_2 \subseteq \ldots$, every k-tangle in $\vec{S}$ induces ℓ-tangles in $\vec{S}$ for all $\ell < k$. If the order function is chosen well, this hierarchy of tangles (see Section 7.4) will correspond to the inclusion hierarchy between species, subspecies, individual organisms and so on: a k-tangle representing an organism will induce the ℓ-tangle of its species etc. This lower-order tangle leaves unspecified some potential features that vary amongst organisms belonging to this species. These potential features will have order at least ℓ and can serve to distinguish organisms of the species defined by its ℓ-tangle.

As pointed out in Section 7.4, designing an order function by hand requires care, and should therefore be left to the experts. At a more technical level, explicitly defined order functions will not normally be submodular. However we can achieve the required degree of submodularity, not of the order function but of the features systems $\vec{S_k}$ they give rise to, by adding corners in $\vec{S_k}$ as needed by our tree-of-tangles algorithm on the fly.

If we do obtain a tree of tangles T as in Theorem 2, one that can distinguish tangles in $\vec{S}$ of different order, this translates directly to a phylogenetic tree for the species (say) that defined the tangles living at the extreme locations of T, its 'leaves' if we view it as a graph-theoretical tree (see Section 8.1). The elements of T, of course, will not in general be single elements of S, but Boolean expressions of elements of $\vec{S}$ that we added as corners in the tree-of-tangles algorithm. So the distinguishers in our phylogenetic tree will not correspond to single DNA positions but to sets of such positions, as is to be expected.

So far, this is just some phylogenetic tree for the set of species represented in V. It will be the Darwinian tree of life, however, if our order function assigned lower order to those DNA positions whose mutations caused earlier branchings in the Darwinian tree.

When we look for some particular DNA in our example, such as in a forensic context, we can exclude other DNA by defining the set $\mathcal{F}$ for which our tangles are $\mathcal{F}$-tangles. The expected or desired outcome of our analysis may then be that there are no tangles in our sample. This can then be documented by Theorem 3, which returns for such a choice of $\mathcal{F}$ an S-tree over $\mathcal{F}$ that witnesses that no $\mathcal{F}$-tangle of S exists.

Let us now turn to a situation in which the molecules in our DNA sample V cannot be reliably aligned. The standard approach in this context is to try to build a phylogenetic tree by clustering the elements of V with respect to some similarity function. DNA molecules from the same organism should form a small and particularly dense cluster, clusters of organisms from the same species should then combine to larger clusters, and so on. These clusters will be nested, so the partitions of V which they define by pitching one cluster, at any level in the hierarchy, against the rest of V form a tree set. Viewed as graph-theoretical tree, this has a good claim to be the Darwinian phylogenetic tree.[6]

The similarity functions on which this clustering approach is based are not unlike those we discussed in Chapter 9. For example, two elements of V are considered as similar if they share many features from a particular list of potential features we deem important. Such a feature might be one of containing a particular substring of bases, or a particular aspect of the corresponding organism's phenotype (such as being able to fly or lay eggs), or of responding in a particular way to some chemical test. But also entropy-based similarity functions have been considered; see [36] for a survey of alignment-free DNA clustering.

We could apply tangles in this approach by using them to compute such a hierarchy of clusters first, as discussed in Section 6.1, and then translate this hierarchy into the desired phylogenetic tree as indicated. However, the detour via clusters is quite unnecessary. If those clusters are induced by tangles (which they witness and guide; see Section 6.1), the tree set defined by the partitions of V that divide those clusters as indicated will be none other than the tree of the tangles that gave rise to those clusters! But that tree set can be computed directly from the tangles themselves, with no need to find clusters that guide them first.

Indeed, this is what motivated the very definition of tangles: the sets of molecules from a particular organism, or from organisms of some common species, will in practice always be fuzzy. Tangles offer a precise way to capture these organisms or species without trying to nail them down as subsets of V. The tree of tangles offered by Theorem 2 will still be a concrete tree set of partitions of V that translates directly to the desired phylogenetic tree: the fuzziness at the level of V no longer exists at the level of relating species, subspecies etc., which are captured as locations – or 'nodes' when we view T as a graph – in our tree of tangles T. More details on this can be found in [12, Section 5.4].

12.4 Drug development: targeting tangles of pathogens

The task we considered in Section 4.3 was to develop a limited number of drugs to combat some large set of different pathogens. The aim would be that each of the drugs we develop can kill many of those pathogens, so that just a few drugs are needed to cover them all. At the same time, the drugs should have only limited side effects. In particular, they should kill only the pathogens they target: not some other, beneficial, organisms we are aware of that are susceptible to the same drugs.

A naive approach would be to consider the set $V = U \cup W$ of both our pathogens and the beneficial organisms we do not want to harm, say with W containing the pathogens and U the beneficial organisms, to somehow divide W into groups of similar kinds of pathogens, and then to develop a drug for each of those groups that kills the pathogens in that group but none of the beneficial organisms in U.

This approach requires that we can identify in W such groups of pathogens of similar kind, which is a typical clustering problem. To approach it in the traditional way, we would decide first when two pathogens are deemed to be similar, and then to seek to find groups in W of pathogens that are similar in this sense. Note that 'similar' is not well defined here even in an abstract sense.[7] But that is a generic clustering problem which need not concern us here. Tangles do offer a contribution to generic clustering, which is discussed in Section 6.1, and again in Chapter 14.

The second problem that comes with this clustering approach is what to do with the clusters once we have found them: how do we train a drug on some given set of pathogens, even if we know that its elements are somehow similar? A general answer to this is that drugs are developed to target pathogens in terms of their features: one drug may target viruses of some particular shape, another those that have some special chemical compound they can bond to. So for this approach to work, those clusters in W, as well as the set $U \subseteq V$ of organisms to be spared, ought to be defined not as arbitrary subsets, but in terms of features of the organisms in V.

But if we intend to follow Section 6.1 and use tangles to identify those clusters in W, then these will be tangles of some set S of potential features of the organisms in W, or more generally of the organisms in V. That is to say, those clusters will be determined not directly as sets of elements of V, but in terms of features of those elements: that is what tangles are.

So our approach is to define clusters in W in terms of features of the organisms in V, because this is how similarity between organisms is defined, and then to capture those clusters in terms of tangles of S, a set of potential features of the organisms in V, because that is how tangles are defined. But then those clusters in W themselves are just a red herring: we can, and should, relate tangles to drugs directly, since both are – or are developed for – sets of features of the organisms in V!

We have discussed extensively in this book how to find tangles in a given set $\vec{S}$ of features. So let us look at the other end now: how should we aim to develop a drug for a given set of features, those forming a tangle of S, so that this drug works as desired?

Let us formalize things a bit more to discuss this properly. Since the potential features of organisms that are relevant for drug development are defined as meaningful properties of these organisms, not merely as partitions of V, every $s \in S$ has a default specification: that which is the intended feature, such as a particular molecular shape, rather than its absence. Let us denote those specifications by forward arrows: let $\vec{s}$ denote the presence of the potential feature $s \in S$, and $\overleftarrow{s}$ its absence. Every tangle of S can then be identified by the subset of those $s \in S$ which it orients as $\vec{s}$: as a collection of certain features from the large pool S of potential features. Given a tangle τ of S, let us write τ^+ for this subset, putting

$$\tau^+ := \{\, s \in \vec{S} : \tau(s) = \vec{s} \,\} \quad \text{and} \quad \vec{\tau}^+ := \{\, \vec{s} \mid s \in \tau^+ \,\}.$$

Our approach is to develop for suitable tangles τ of S a corresponding drug D_τ. All we know at this stage is that this is prima facie a sensible approach, simply because both tangles of S and the drugs we seek to develop are defined in terms of features in $\vec{S}$. But there is also one obvious way to establish a correspondence between the two: to design D_τ so as to target, somehow, organisms with features in $\vec{\tau}^+$. Let us see whether that is a sensible aim, and if so how it might be pursued.

Our drugs D_τ should meet certain requirements. Ideally, if $V(D)$ denotes the set of organisms in V killed by the drug D, we would want for all τ that

(1) $V(D_\tau) \subseteq W$; and that

(2) $V(D_\tau)$ is large enough to make the development of D_τ worthwhile.

In addition, every $v \in W$ should ideally lie in $V(D_\tau)$ for some tangle τ.

The simplest way to develop D_τ with reference to τ^+ would be that it should kill any organism that has one of the features in $\vec{\tau}^+$. Expressing features as subsets of V, as usual,[8] this can be expressed as aiming for $V(D_\tau) = \bigcup \vec{\tau}^+$. But this set of organisms is way to large to satisfy (1).

At the other extreme we might go for $V(D_\tau) = \bigcap \vec{\tau}^+$, so that D_τ kills only organisms that have *all* the features in $\vec{\tau}^+$. But this set of organisms is likely too small to satisfy (2), at least as soon as S is large enough that $\bigcap \vec{\tau}^+$ can satisfy (1).

The dilemma between those two options is exactly what tangles are designed to resolve.[9] Neither a simple union nor a simple intersection is the right way to associate with τ^+ a target set of organisms for D_τ. But just as the task of developing a good drug from knowing τ^+ is more complex than targeting either each of the features in $\vec{\tau}^+$ separately, or only all of them at once, what makes the set $\vec{\tau}^+$ of features into a tangle is also more subtle than just taking its union or intersection.

Our approach, therefore, is to use in the development of D_τ the tangle properties of τ, including its order under a suitable order function and its agreement value, rather than just the set τ^+ or $\vec{\tau}^+$. How exactly this might be done, however, is not a question we can answer once and for all here: it is intricately bound up with our choices for U, W, and S, and will depend on those.

Yet even at this general level there is an intriguing coincidence between some sensible relationship of τ^+ and $V(D_\tau)$ on the one hand and a corresponding relationship in tangles on the other. Not for tangles of S, though, but for tangles of its dual feature system!

This possibly sensible relationship between τ^+ and the set $V(D_\tau)$ of organisms for D_τ to target is the following compromise between the two extremes of aiming for either $\bigcup \vec{\tau}^+$ or $\bigcap \vec{\tau}^+$ as $V(D_\tau)$. Let us instead try to develop D_τ so that it kills those $v \in V$ that have neither at least one, nor all, but at least some minimum proportion p of the features in $\vec{\tau}^+$, for some $p \in (\frac{1}{2}, 1]$.

This is reminiscent of guiding sets for tangles, and there is indeed a connection. Recall that, for the feature system $\vec{V}$ dual to $\vec{S}$ as defined in Section 7.7, our default orientations in $\vec{S}$ define default orientations

$$\vec{v} = \{\, s \in S \mid v \in \vec{s}\,\}$$

for all $v \in V$. Our choice of $V(D_\tau)$ as the set of those $v \in V$ that have at least $p\,|\tau^+|$ of the features in $\vec{\tau}^+$ can thus be rewritten as setting

$$V(D_\tau) := \{\, v \in V : |\vec{v} \cap \tau^+| \geqslant p\,|\tau^+|\,\}.$$

Now this is, in essence, nothing but an orientation of V guided by τ^+ with reliability at least p: the set $\tau^+ \subseteq S$ guides the specification of v as $\vec{v}$ for precisely those $v \in V$ that lie in $V(D_\tau)$.[10] Note that for this relationship to hold we do not need that the orientation of $V(D_\tau)$ guided by τ^+ is a tangle, let alone extends to a tangle of all of V.

A potential application of tangles to drug development based on this generic choice of $V(D_\tau)$ might thus consist of the following steps. Let V be a set of organisms made up of a pool W of pathogens for which we wish to develop some drugs, and a set U of beneficial organisms on which we wish to test those drugs for side effects. Develop a list S of features that may be found in the pathogens in W and which can, in groups, be targeted by some common drug to be developed for this group. Use tangle software to identify specifications O of the dual feature system $\vec{V}$ whose positively specified subset $O \cap \{\, \vec{v} \mid v \in V\}$ has a guiding set F in S. Its underlying subset $O^+ := \{\, v \in V \mid O(v) = \vec{v}\,\}$ of V consists of those organisms in V that have most of the features in F. If all of O is guided by F then, conversely, also every $v \in V \smallsetminus O^+$ lacks most of the features in F.[11] Discard O unless O^+ consists of pathogens only. If it does, develop a drug, specifically for this collection F of features, that kills any organisms that has most of the features in F while leaving all others unharmed. This drug will then work on all the elements of O^+, which by assumption lie in W, while leaving unharmed all the elements of $V \smallsetminus O^+$, which include the beneficial organisms in U.

13

Applying tangles in the social sciences

In this chapter we revisit the scenarios discussed in Chapter 5 of how tangles might typically be applied in the social sciences. An exception is Section 5.3, in which we saw how tangles can quantify Wittgenstein's Family Resemblances. We shall not revisit that section here, but take it up again in Chapter 14 from an AI computing perspective.

Our focus in all the examples revisited here will lie on the role played by the tangle theorems from Chapter 8, and in particular on the role played by the tangle-distinguishing tree sets provided by Theorems 1 and 2. To avoid repetition, we shall discuss aspects that apply in most or all of these contexts only once, for the mindset example in Section 13.1. This section, therefore, is recommended reading also for readers primarily interested in another field discussed later.

13.1 Sociology: from mindsets to matchmaking

In the scenario discussed in Section 5.1 we had a questionnaire S, which some people in a set V have filled in and returned to us. We may think of V as a sample from a larger population P of people in whose views we are interested. The sample V should then be chosen so as to represent P well.

A tangle of S, which we call a *mindset* in this context, is a typical pattern of views of the individuals in V. This pattern may have been previously known, or suspected to exist, or it may be a newly discovered pattern. In fact, while we are usually interested in understanding mindsets, found as tangles, in terms of some intuitive interpretation, the theory does not require us to find interpretations. In particular, we can use it to make tangle-based predictions for the views or behaviour of people in $P \smallsetminus V$ without such an intuitive understanding; see Section 12.1.

Theorem 1, our tree-of-tangles theorem for fixed S, finds a tree set T of questions that are critical for telling apart the mindsets found as tangles. Both the tangles found and the tree set T that distinguishes them will typically contain not just elements of S but also Boolean expressions of elements of $\vec{S}$: logical combinations of views on questions in S. Compare Section 11.3.

The fact that T is a tree set is likely to correspond to logical inference. For example, if S is a questionnaire about various types of sport, it might have one tangle representing enthusiasm for skiing, and another for boxing. Since these are distinct mindsets, T will have to contain a question that tells them apart. If we assume that no-one likes both skiing and boxing, this might be the question s asking 'do you like skiing', or the question r asking 'do you like boxing'. If the yes-answers to these questions are $\vec{s} = A \subseteq V$ and $\vec{r} = B$, respectively, and we know that $A \cap B = \emptyset$, we would have $A \subseteq \overline{B}$ and $B \subseteq \overline{A}$, respectively.

More generally, any two elements r, s of T will have specifications $\vec{r}$ and $\vec{s}$ that are disjoint as subsets of V: this is simply a consequence of the fact that they are nested. This will happen if $\vec{r}$ and $\vec{s}$ are contradictory statements, assuming the answers of each individual in V are logically consistent. The converse implication, however, need not hold: we have $\vec{r} \cap \vec{s} = \emptyset$ as soon as no individual in V holds both views, $\vec{r}$ and $\vec{s}$. This can happen also when $\vec{r}$ and $\vec{s}$ are logically consistent but V is not large enough to represent the combined view $\vec{r} \wedge \vec{s}$.

Suppose now that there is also a mindset that corresponds to a loathing for all sports. Then T must contain questions that distinguish this mindset from each of our two sport-loving ones. Neither r nor s achieves both these aims at once: r fails to distinguish the sport loathers from the skiers (since these also dislike boxing), while s fails to distinguish them from the boxers. If T achieves the required distinctions by the question t asking 'do you like any sport', with yes-answer $\vec{t} = C$ say, then the nestedness of r, s and t will play out as $C \supseteq A$ and $C \supseteq B$, which reflects logical inference.

The locations[1] of T correspond to the various mindsets that T distinguishes. If T reflects logical inference, then each location consists of the strongest views in $\vec{T}$ that are consistent with the mindset at that location. It thus represent this mindset 'in a nutshell'.

In our earlier example, if we have just one further mindset, that of loving sports in general but hating both skiing and boxing, the location of this mindset would be the star $\{\vec{t}, \overleftarrow{s}, \overleftarrow{r}\}$ in $T = \{r, s, t\}$, while the location of the mindset of loathing all sports would be just $\{\overleftarrow{t}\}$. If there is no such fourth mindset, then T need not contain all three questions r, s, t to distinguish the three mindsets of skiing, boxing, and loathing all sports. It might contain t and s, say, in which case $\{\vec{t}, \overleftarrow{s}\}$ would be the location of the mindset of loving sports in general, but not skiing. This mindset would differ only immaterially from the boxing mindset, since there are not enough sports enthusiasts that dislike both skiing and boxing to form their own mindset. Similarly, T might contain t and r but not s, since the skiing fans are indistinguishable from the general sport fans that dislike boxing.

In Theorem 3, the locations σ of T are in $\mathcal{F}$. If $\mathcal{F} = \mathcal{F}_n$ for some n, these are sets of views held by only a few people in V, fewer than n. If we hand-designed $\mathcal{F}$ to make it contain also some sets σ of views explicitly, the people sharing the views in such a set σ might be numerous (so that $\sigma \in \mathcal{F} \setminus \mathcal{F}_n$) but deemed irrelevant to our survey S. The entire set V is then fragmented by T into these small or irrelevant subsets, which documents conclusively that no significant mindset in terms of S exists in the population V – and, by inference, in the population P.

In a more subtle setup we might have an order function on S that assigns low order to questions of a fundamental or basic nature, and higher order to questions about more detailed aspects of mindsets expressable in terms of those basic views. Then the views held by a mindset τ of S_k refine those of the mindset $\tau \cap \vec{S_\ell}$ of S_ℓ for each $\ell < k$.

In our earlier example, the question t about sports in general would have lower order than the questions r and s about boxing and skiing, since it is less specific. We might then have a tangle of low order for sports enthusiasm, and tangles of higher order for the love of skiing and boxing. These latter tangles would both refine the general sports tangle. In particular, all three would exist as separate tangles even if, as in our earlier example, there are not enough people that love sports in general but hate both boxing and skiing to form their own mindset.

The locations of the ℓ-tangles in $\vec{S}$, with ℓ small, in the tree set $T \cap S_\ell$ found by Theorem 2 are the collections of strongest consistent views in $\vec{T}$ for the mindsets at this basic level ℓ. As detailed in Section 7.4, these mindsets then split into more refined mindsets, k-tangles in $\vec{S}$ with $k > \ell$, whose locations in $T \cap S_k$ are sets of strongest consistent views in $\vec{T} \cap \vec{S_k}$.

Theorem 2 thus enables us to extract, from the data provided by S, a hierarchy of mindsets of varying degree of complexity, which unfold from the most basic to the most specific. We also have in T a small selection of key questions on which these mindsets differ, so that we can tell them apart just by evaluating those few questions.[2]

Finally, we have the option of not choosing the order of the questions in S explicitly, but to compute orders of all partitions of V as outlined in Chapter 9, and then to consider for Theorem 2 the sets $S_1 \subseteq S_2 \subseteq \ldots$ arising from the order function chosen. This approach has the advantage that it is entirely deterministic once the questionnaire S has been set up, in that we do not interfere with the result – what mindsets are found as tangles – by declaring some questions as more basic, fundamental, or important than others. On the other hand, we have less influence on what kinds of possible mindsets are found.

Once we have determined the mindsets in terms of S that are prevalent in our population V we can, if desired, associate individuals $v \in V$ with mindsets that represent their views. As discussed in Section 6.1, there are various ways of doing that, and which is best will depend on the context. Note, however, that the simplest way of associating an individual with a mindset representing them, which is to assign a person v to the tangle τ_v^+ of S closest to his or her entire set $v(S)$ of answers to S, is not available for the larger population P whose views are known only for the questions in T. For these individuals, the tangle of S that represents their views best is likely to remain the unique tangle of S that extends their views $v(T)$ on T, assuming that these are consistent. This is discussed further in Section 14.1.

To illustrate all this by a concrete example, let us apply our tangle-based approach to detecting mindsets to design a matchmaking algorithm for a dating app. Let us assume, for simplicity, that our brief is to match individuals of similar character, say from some population P. Then S will be be a questionnaire of character-related questions. We assume that some significant set V of people have already answered S, and on the basis of this data we hope to match new users $p, q \in P$.

Any tangle of S is a complete set of answers to the questions in S that are typically found together. We may think of each tangle as representing a particular type of character found amongst the people in V. Our tangle algorithms will compute the most prominent such character types, perhaps with a hierarchy of subtypes as outlined in Section 7.4.

For our matchmaking algorithm, however, it is not enough to know these types, since we have to match individuals. Can our knowledge of character types, encoded as tangles, help us do this in a way different from, and perhaps better than, just pairing individuals who agree on the greatest number of questions?

One way in which this might be possible is that our tangle analysis of how the people in V answered S identifies a few particularly significant questions: questions such that, for any two concrete individuals $p, q \in P$, agreeing on those questions is a stronger indicator of like character than agreeing on some of the other questions.

Theorems 1 and 2 indeed provide us with exactly this: the tree set T they output is a small, if desired even minimal, set of questions in S by which any two typical ways of answering S can be distinguished. When we compare the answer sets $p(S)$ and $q(S)$ of our two individuals $p, q \in P$ by counting the number of questions on which they agree, we can give questions in T larger weight than the other questions in S. In the case of Theorem 2 we can even differentiate within T by giving questions of lower order greater weight than questions of higher order, because they distinguish tangles representing more fundamental character types than questions of higher order do. We might even follow the approach of Section 12.1 and interview the individuals in P, those using our app, only on the questions in T.

If p and q answer the questions in T consistently in the logical sense, as we shall assume, and our data pool $V(S) = \{ v(S) \mid v \in V \}$ is large enough to reflect logical consistency in $\vec{T}$ by consistency as we defined it extensionally in terms of $V(S)$, then the specifications $p(T)$ and $q(T)$ of T will be consistent and thus define tangles of T, in the case of Theorem 1, or k-tangles in $\vec{T}$ in the case of Theorem 2. We might then define the 'matching score' of p and q as

$$ m_T(p, q) := \sum \{\, 1/|t| : t \in T \text{ and } p(t) = q(t) \}, $$

a weighted count of the $t \in T$ on which p and q agree that gives greater emphasis to more fundamental questions, those of lower order.[3]

With this approach, two individuals p and q are likely to be matched if they share the most fundamental ones amongst those traits by which different character types are most clearly distinguished, the traits tested by the questions in T, and differ only one some more specific traits that distinguish subtypes of character.

This method ignores how p and q answered the questions in $S \setminus T$, those that in our tangle analysis of $V(S)$ we found to be less relevant to distinguishing types of character, and hence they need not be tested on those questions. If p and q are asked to answer those questions as well, their matching score $m_T(p, q)$ could be complemented by a simple count of how many of those questions they agree on in addition.

13.2 Psychology: diagnostics, new syndromes, and the use of duality

In Chapter 5.2 we discussed two possible applications of tangles in psychology.

The first was the formation of non-standard concepts[4] from a plethora of perceptions in the minds of patients with a similar psychiatric condition: perceptions that combine loosely into a notion in these patients' minds which cannot be naively understood by the therapist, because these perceptions are not commonly associated with each other to form a concept in our minds. The value of tangles in this application lies in finding them, and thereby in identifying such unfamiliar notions.

Since every tangle in this setup will be a concrete list of perceptions, therapists can then train their own intuition to convert these lists into something like notions for themselves too: notions that are new to them but to which they can assimilate, in an effort to better understand their patients. The tangle theory we developed in Part III does not really add to this, so we shall not return to this tangle application here.

Our other application of tangles from Section 5.2, however, has a larger scale and goes beyond just identifying and understanding tangles. The idea there was to see which psychological symptoms, collected across many patients, combine into recognizable syndromes that can serve as a focus for more targeted research and the development of treatments.

An obvious possible direction for a large and fundamental study as envisaged in Section 5.2 would lie on discovering previously unknown syndromes. However, any tangles-based definition of 'syndrome' will also encompass known syndromes, which should likewise show up in such a study. This can be used in two ways. First, to validate the tangle-based model by checking that known syndromes and psychological conditions are reliably found. Secondly, tangles will offer some quantitative underpinnings also to the ongoing study of previously known syndromes.

The set T of tangle-distinguishing features found by Theorems 1 and 2 on the basis of such a large study can then be used also for diagnostic purposes in individual patients. As discussed in Section 12.1, the idea would be to test T, rather than the much larger S, on a given patient first, and if this returns a consistent specification of T we will have identified a tangle that is most likely to describe the patient's psychological condition: the unique tangle of S that extends this consistent specification of T.

On the face of it, this procedure resembles the familiar classical diagnostic approach in medicine based on 'exclusions': just as a doctor tests specific conditions in order to exclude potential illnesses as the cause of the symptoms observed, testing each potential feature in T will support some and exclude other possible tangles of S. Indeed, T may well be seen as a particularly efficient set of such tests, which evidence-based guidelines to doctors may suggest.[5]

However we must not forget that each tangle, being a particular specification of S, is just an 'ideal' set of test results that corresponds to a syndrome. Diagnosing an individual patient with one of these syndromes is still a nontrivial process: we have to find the syndromes that best match this patient's own test results on S. There are several ways to do this – see Sections 6.1, 14.1, and 14.2 – and this should be a matter for learned debate within the discipline as much as for the psychologist in charge of the patient.

Finally, there is an aspect of tangle duality that is particularly useful in a psychology context. Let us again think of S as a set of questions answered by every person v in V, even though in our current context the $\vec{s} \in \vec{S}$ are more likely observations made by a psychologist than answers given directly by the client.

Recall from Section 2.5 that a set $X \subseteq V$ *guides* a specification τ of S, not necessarily a tangle, if for every $s \in S$ the majority of the elements of X specify s as τ does, rather than in the opposite way. If this majority subset of X has size at least $p\,|X|$ for every s, and $p \in (\frac{1}{2}, 1]$ is maximum with this property, then X guides τ with *reliability p*.

Let us call such a set X, one that guides some specification of S with reliability at least p, a *p-decider* for S. Thus, X is a p-decider for S as soon as for every $s \in S$ at least $p\,|X|$ of the elements of X specify s in the same way. If $\frac{1}{2} < p \leqslant \frac{2}{3}$, the specification of S defined by X need not be a tangle; if $p > \frac{2}{3}$ it will be. If p is close to 1, the elements of X are essentially indistinguishable in terms of S: every question in S

is answered by most of them in the same way.

Now recall from Section 7.7 that the elements $v \in V$ of the ground set of our feature system $\vec{S}$ are themselves potential features in a dual feature system $\vec{V}$ with ground set S. A p-decider for the dual system $\vec{V}$, then, is a subset S' of S such that every $v \in V$ answers most of the questions in S', at least $p|S'|$ many, in the same way.

The questions in S', therefore, can be used interchangeably to some degree, which can be particularly useful in a psychology context. For example, if we are interested in how our patient would answer some particular question $s' \in S'$, but we cannot ask s' directly because of some taboos or traumas, or because asking s' might influence his or her other answers, we can instead ask some more innocuous $s'' \in S'$ and interpret its answer as an answer to s'.

There is still a small catch with this. If some p-decider S' is a little better than required, in that every $v \in V$ answers at least $p'|S'|$ of the questions in S' in the same way for some p' a little bigger than p, we can add any $s \in S \smallsetminus S'$ to S', no matter how different from the existing $s' \in S'$, and the resulting set $S' \cup \{s\}$ will still be a p-decider for $\vec{V}$. We shall not want to replace a question $s' \in S'$ with this s. Hence in order to make such replacements in practice, we can look for them within p-deciders, but have to remain vigilant when picking a concrete replacement.

Alternatively, we can use tangles of partitions of S to determine subsets S' that form more traditional clusters in S than p-deciders for $\vec{V}$ are, and choose our replacements within these clusters. See Sections 6.1, 14.1, and 14.2 for how to do this. An example, where tangle-based clusters in S were used to analyse the 'Big Five' approach to personality tests, can be found in [3].

13.3 Political science: appointing representative bodies

The starting point for this example was that bodies elected by some population to make decisions on their behalf should be composed of delegates that represent the typical views held in the electorate, perhaps in numbers reflecting the popularity of these views. Given that tangles provide a model for precisely this, views commonly held together, one might seek to constitute a representative body by appointing as delegates people whose views best resemble the tangles found for the set S of relevant issues determining the election.

In a first approach, we might delegate to our representative body those $v \in V$ whose views are closest to the various mindsets found by S. For each tangle τ of S we might appoint one representative, the $v \in V$ for whom $|\{\, s \in S : v(S) = \tau(s)\}|$ is maximum amongst all the people in V.

But even if all the main views held in the electorate V are now represented in our representative body, the question is still whether that representation is fair: more commonly held views should be represented by more delegates.

Theorem 2 offers a neat way to do this. Recall that our algorithm which computes the tangles for a hierarchy $S_1 \subseteq S_2 \subseteq \ldots \subseteq S$ of feature sets does so, in its simplest form described in Section 11.2, for the sets S_k in turn: as $k = 1, 2, \ldots$ increases. At each stage k it evaluates whether or not a tangle found for S_k extends to one or several tangles of S_{k+1}. At this time, the algorithm can already compute the subset T_k of T needed to distinguish all the tangles in $\vec{S}$ of order up to k. The tree set T_k partitions V into the sets $V_\tau := \bigcap\{\, A \mid A \in \tau \cap \vec{T_k}\}$, where τ varies over the consistent specifications – or tangles – of T_k.

At this stage we could establish our representative body by appointing for each τ a number d_τ of delegates proportional to the number $|V_\tau|$ of people whose views are best represented by τ.

However, Theorem 2 enables us to improve on this further, by choosing the composition of the group of delegates for τ not arbitrarily (specifying only its size) but according to the tangles of higher order into which τ splits: the tangles of $S_{k+1}, S_{k+2}, \ldots$ that induce τ. In other words, when we run our algorithm we make a decision at stage k for every τ as above whether or not we seek to refine T_k at the node representing τ: for those τ whose $|V_\tau|$ is smallest we appoint one delegate, while for the τ with bigger $|V_\tau|$ we keep the algorithm running. We can do that so that the tangles τ at which we stop the algorithm will have varying order, but they will represent the views of roughly the same number of people in V. So it will be fair to delegate one person from each of those sets V_τ to our representative body.

Theoretically, even this refined process can return sets V_τ that are substantially bigger than others: it can happen that a tangle τ is represented by many people but still does not split into higher-order tangles. Then we should still pick different numbers d_τ of representatives for different tangles τ. But this procedure uses such arbitrary choices of representatives only as a last resort, when there are no alternative choices informed by higher-order tangles.

Unlike in a party-based representative vote, the composition of delegates obtained in this way reflects not only the relative support of the major views in the electorate. It also reflects within each of these views its various shades of more refined views, all proportionately to how these views are represented in the electorate.

13.4 Education: devising methods and assigning students

In Chapter 5, Section 5.5, we discussed how tangles can help us to group teaching techniques into 'methods' so that different types of students can benefit best from techniques that work particularly well for them. A 'method' in this sense was simply a collection of techniques that are applied at the same time in one class. The idea is that, since at a large school it may be possible or necessary to offer the same content in several parallel classes, these classes could use different such 'methods' to benefit different types of students.

Our approach was to define 'methods' as tangles of techniques: groups of techniques that often work well together, where 'often' means 'for many potential students'.

Finding these tangles will require a fundamental study involving a large set V of students. Every student in such a study will have to be tested against various techniques, say those in a set S, which requires a considerable effort over time.[6] Once these tests have been performed we can define an order function on S to distinguish techniques that made a big impact (positive or negative) on many of the students tested: such techniques will be assigned a low order to ensure that they are considered even for the most basic tangles computed.

The k-tangles found will then be concrete specifications of some techniques, ranging from the fundamental to the more subtle as k increases. They each form a 'method' in that they lay down which of the techniques they specify should be used or not when teaching 'according to' that tangle.

All this cannot be done at a single school facing the practical problem of how to set up its classes and assign its students to these classes. Conversely, not every 'solution' found by a large study, such as a set of tangles and a tangle-distinguishing set T of critical techniques, can be implemented at any given school. Indeed, local requirements or possibilities may dictate how many classes can be taught in parallel (and hence, into many tangles or 'methods' the techniques available should be grouped), and which techniques are available may depend on the composition of the local teaching staff.

However, once we have the data from our large study, a complete specification $v(S)$ of S for every student v tested, we can run our tangle algorithms with different parameters, parameters than can be adjusted to a school's given requirements or possibilities. For example, we can adjust the agreement parameter n in the definition of $\mathcal{F}_n$ to obtain a desired number of tangles to match the target number of parallel classes, or we can delete potential techniques s from S before computing the tangles if these techniques cannot be implemented at the given school.

The result of this process, then, is a tangle-based concrete way to group the techniques available into as many 'methods' as parallel classes can be offered.

But how do we assign students to these classes? One possibility is to simply describe the methods on which the various parallel classes are based, and let them choose. Another is to test, or at least ask, students which techniques from the specific set T found by the algorithm for Theorem 1 works best for them. By definition, this is a small set of 'critical' techniques, each a combination of the techniques tested directly (those in S), on which the methods found differ. If T was chosen minimal, then every consistent specification of T defines a unique tangle that extends it – one of the teaching methods that correspond to the classes offered in parallel.

As with our earlier task of determining a set of methods to be offered, the set T that critically distinguishes them can be computed locally to match the specific school's possibilities and requirements. In this way, the results of the same large study can be used by different schools to compute a bespoke set of particularly critical teaching techniques against which students can be tested in order to assign them to the classes corresponding to the various methods.

As the number of tangles involved in these applications will be small, the way to do this will be to assign each student v to the class whose tangle τ extends his or her specification $v(T)$ of T, the class $V_\tau := \bigcap \{\, A \mid A \in \tau \cap \vec{T} \,\}$ defined after Theorem 1 in Section 8.2.

14

Applying tangles in data science

As we have seen, tangles of a set S of potential features of a set V of objects are best at discovering what groupings – such as mindsets – exist in V, rather than at dividing V into such groups. However, once found, tangles of S can help with this too. In the first two sections of this chapter we revisit this matter, building on our informal discussion in Section 6.1.

In the next three sections we address some questions that data scientists with a clustering background might have when they first hear about tangles. Section 14.3 highlights the difference between the tangles of a set S of potential features (of the points in a given set V) and traditional clusters in the set $\vec{S}$ of the corresponding features, which is sometimes called the 'feature space' considered for the elements of V. In Section 14.4 we contrast our tangle-finding algorithm, which chooses a sequence of partitions s of V and focusses after choosing s on the tangles that live on[1] either side of s, with standard hierarchical search and clustering in V, which superficially does something quite similar. In Section 14.5, finally, we compare tangles of feature systems defined by spectral methods (Section 10.2) with traditional spectral clustering.

In the remaining four sections we revisit the examples of potential tangle applications that we discussed informally in Chapter 6: applications to image segmentation and compression; to classifying texts; and to teaching computers meaning. The chapter winds up with a concrete and fun project in this latter area: how we might use tangles to set up an interactive thesaurus.

14.1 Clustering by tangles: hard clusters or none

Let us recall from Section 1.3 the motivation for the notion of tangles from a clustering perspective. The starting point was the truism that clusters in a given dataset V, however they might be formally defined, would not be divided roughly in half by partitions of V at its 'bottlenecks': places where V was 'narrower' than any cluster. Each cluster would therefore orient all such 'bottleneck partitions' towards the side on which most of it lies. These orientations of all the various bottleneck partitions by any one cluster would be consistent in that they would all point in a common direction: towards that cluster. The set of all these consistently oriented bottleneck partitions, to be called a 'tangle', would then be something like a structural footprint of the cluster. It would be precisely defined even if the cluster itself might not be.

The appeal of this idea lay in that, unlike with clusters defined, somehow, directly as subsets of V, it would relieve us of having to decide for each $v \in V$ to which cluster, if any, it should be assigned. Tangles would tell us 'where' the clusters in V were, not 'what' they were. A further promise was that the structural information retained in tangles from the clusters that defined them would be something like their essence: enough information to enable us to prove the main structural theorems about point clusters once they were properly defined, but independently of how exactly this was done.

If such structural information was not deemed to be enough, however, but some interest remained in having concrete clusters defined as point sets, we might still look at how such traditional clusters might be retrieved from our tangles once those had been determined.

When this idea behind tangles was outlined in Section 1.3, our preliminary notions of bottlenecks and of consistency, both needed for the notion of a tangle, required that we already had an informal idea of what the clusters in our dataset V were: its 'bottleneck partitions' were *defined* (provisionally) as those partitions of V that would not cut right through any cluster, and their 'consistent' orientations were *defined* (provisionally) as those towards some fixed cluster. The radically different idea of *basing* a structural analysis of our dataset V on tangles, rather than on traditionally defined point clusters, therefore required that we first came up with definitions of bottleneck partitions and of consistency that were independent of any pre-conceived notion of point clusters.

In Chapter 7 we gave such an independent definition of the tangles of a given set S of partitions. In Chapter 9 we studied various ways of

determining which partitions might count as the 'bottleneck partitions'
of V forming such a set S, which would thus be the raw material for
our tangles. In Chapter 8 we saw that our notion of tangles supports
substantial structural analysis.

Let us get back, then, to our original clustering question, now that
we have tangles up and running: the question of how to associate with
a given tangle a subset of V that might count as a point cluster. When
we addressed this question in Section 6.1, we distinguished three types
of clusters associated with tangles: 'hard', 'soft', and 'fuzzy'.

As candidates for hard clusters associated with a given tangle τ of S
we studied the sets $X_\tau \subseteq V$ of points which, in the way they orient S,
agree best with τ amongst all the tangles of S. Now that we have the
tangle theory from Sections 7 and 8 at our disposal, we can refine this
approach; we shall do this first in this section.

Our tree-of-tangles theorems from Chapter 8 also allow us to associate with our tangles some candidate sets for hard clusters that are quite
different from the sets X_τ considered in Section 6.1. Unlike those, they
will be precisely delineated from the outset, by the partitions in the tree
of tangles: every $v \in V$ will belong to exactly one of them, not just 'by
degree' as in the case of the X_τ. As a consequence, these 'hard clusters'
do not give rise to corresponding soft clusters in the way the X_τ do.
Describing these sets will make up most of this section.

Finally, we shall apply the tangle–tree duality theorem to obtain
easily checkable witnesses to the non-existence of tangles, and thereby
of any tangle-induced clusters.

Let us start by refining our candidate sets X_τ for clusters to be
associated with a tangle τ of a given feature system $\vec{S}$. Our definition
of X_τ in Section 6.1 ensured that the elements of X_τ were 'most like τ'
compared with other tangles, in that

$$\sigma(v, \tau) := |\{\, s \in S : v(s) = \tau(s) \,\}|$$

was maximum for these v amongst all the tangles of S. Let us now, more
broadly, consider any similarity measure between specifications of S:
any function σ that assigns a real number $\sigma(f, g)$ to every unordered
pair $\{f, g\}$ of specifications of S. These f and g could be elements of V,
because every $v \in V$ specifies S, as $\{\, v(s) \mid s \in S \,\}$. They could also be
any other orientations of S, which may or may not be tangles. Setting

$$\sigma(f, g) := |\{\, s \in S : f(s) = g(s) \,\}|$$

for any two specifications f and g of S is one option, but not the only one.

To refine our sets X_τ from Section 6.1, let us now consider for τ not just tangles of the entire set S but k-tangles in $\vec{S}$, i.e., tangles of the sets $S_k = \{ s \in S : |s| < k \}$ for $k = 1, 2, \ldots$. Here, $s \mapsto |s|$ can be any order function on S, say one of those discussed in Chapter 9.

As a first approach, we might associate each $v \in V$ with the tangle in $\vec{S}$ of largest order that is a subset of $\tau_v = \{ v(s) \mid s \in S \}$. In practice, we would first look which tangle τ_1 of S_1, if any, is a subset of τ_v; then which tangle τ_2 of S_2 refining τ_1 is still a subset of τ_v; and so on until we have found a tangle $\tau_k \subseteq \tau_v$ such that no tangle of S_{k+1} that refines τ_k is still a subset of τ_v. We would then associate v with this τ_k. Conversely we might, for every tangle τ in $\vec{S}$, redefine X_τ as the set of those $v \in V$ that we associate with τ in this way.

If S is a well-chosen set of balanced 'bottleneck' partitions of V, such as those we looked at in Chapter 10, or S is a well-designed questionnaire, the sets X_τ redefined in this way could make reasonable clusters in V. As originally envisaged in our furniture example in Chapter 2 and detailed in Section 7.4, the sets X_τ would form a hierarchy of subsets that could be considered as clusters at different levels of detail. In particular, not every $v \in V$ has to lie in one of these sets: depending on our choice of $\mathcal{F}$, it may well happen that τ_v contains no $\mathcal{F}$-tangle in $\vec{S}$. In the example, this might reflect the reality of a singular piece of furniture that does not lie in any of the standard categories such as chairs, tables or beds.

If S is larger, however, the above approach to defining sets X_τ associated with the tangles τ in $\vec{S}$ will be too narrow. Assume, for example, that S is the set of all partitions of V. Then $\{v\} \in \tau_v$ for every $v \in V$. If these singleton partitions have minimum order, as is often the case, then any tangle in $\vec{S}$ has to orient them. Hence if $v \in X_\tau$, then $\tau \subseteq \tau_v$ has to contain $\{v\}$ rather than $V \smallsetminus \{v\}$. However this is unlikely to happen, since singletons such as $\{v\}$ do not lie in many tangles of interest, such as $\mathcal{F}_n$-tangles for $n > 1$. And if they lie in none, then $X_\tau = \emptyset$ for all τ.

Figure 14.1 illustrates this. Let S be the set of all partitions of V endowed with the order function (O1), consider $\mathcal{F}_n$-tangles in $\vec{S}$ for some $n > 1$, and let ℓ be the order of the blue partitions. Then no tangle in $\vec{S}$ orients any of the black partitions s outwards, as $\vec{s}$ say: that would force it to contain smaller and smaller subsets of $\vec{s}$ too, and ultimately a singleton $\{v\}$ with $v \in \vec{s}$.[2] But no tangle contains these by our choice of n. So S has only one k-tangle for every $k \leqslant \ell$, which orients all the partitions in S_k inwards; it has two k-tangles, distinguished by t, for $k > \ell$ up to some value of k at which partitions of order k can cut away more than a third of the two point clusters X or Y; and it has no k-tangles for any larger values of k. Compare Section 7.4.

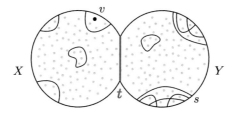

FIGURE 14.1.   Two tangles distinguished only by t.

Our revised definition of X_τ, however, does not reflect this. Rather than returning two sets X_τ for the two k-tangles for $k > \ell$, such as the sets X and Y shown in the picture, we get $X_\tau = \emptyset$ for all tangles τ in $\vec{S}$. This is a consequence of setting $n > 1$ and the fact that the singleton partitions $\{v\}$ have minimum order. But setting $n = 1$ would not help either: the only tangles in $\vec{S}$ would then be the k-tangles τ_v, for any k, so all our 'clusters' X_τ would be of the form $X_\tau = \{v\}$ for some v.

So let us revise our definition of the sets X_τ once more. Let us revert to the original definition of X_τ from Section 6.1, which placed a point v in X_τ if τ was the tangle it agreed with most (rather than completely, as in our requirement that $\tau \subseteq \tau_v$), but let us apply this iteratively for all the k-tangles of S as $k = 1, 2, \ldots$, attaching more importance to those of lower order.

For example, we might measure the similarity of two specifications f, g of subsets of S by a vector

$$\sigma(f, g) = \big(\sigma_1(f, g), \sigma_2(f, g), \ldots \big)$$

where $\sigma_k(f, g)$ for $k = 0, 1, \ldots$ denotes the number of partitions $s \in S$ of order k such that $f(s) = g(s)$ (which is meant to imply that f and g are both defined on s), and order these vectors $\sigma(f, g)$ lexicographically.[3]

Alternatively, we might define the similarity of f and g as

$$\sigma(f, g) := \sum \{ N - |s| : f(s) = g(s) \},$$

where N is any constant large enough to make all these values of $\sigma(f, g)$ non-negative. As in our original definition of X_τ in Section 6.1, this counts the $s \in S$ which f and g specify in the same way; but now elements s of low order are given greater weight in this count. The difference between this and our earlier lexicographic definition of $\sigma(f,g)$ is that if $f(s) = g(s)$ for many s of high order, then this can offset the fact that $f(s) \neq g(s)$ for a few s of low order (and vice versa): low-order partitions no longer override all higher-order ones.

With such a refined similarity measure σ in place, we can simply re-use our original definition of the sets X_τ from Section 6.1: for each v, pick a tangle τ in $\vec{S}$ for which $\sigma(v, \tau)$ is maximum, call it τ_v^+, and let

$$X_\tau := \{ v \in V \mid \tau = \tau_v^+ \}.$$

In the example of Figure 14.1, this will return the two clusters X and Y as sets X_τ associated with the two k-tangles τ for $k > \ell$.

It might seem like a drawback of this refinement that now, once more, every $v \in V$ will come to lie in some X_τ. However those v that we feel do not lie in any cluster, such as the points in the handle in Figure 1.3, will lie in X_τ only for the unique tangles τ of low order. In our examples, these would be the tangles that orient all those low-order partitions of the figure inwards. We may therefore think of such X_τ as a kind of bin that collects all those v that do not lie in any natural cluster. Since our tangle analysis tells us whether some τ is the only tangle of its order or not, we can tell the bin from the clusters, so some v may avoid being assigned to any of our clusters X_τ after all.

Let us now turn to our second topic in this section: a way of associating candidate sets for clusters in V with the tangles in $\vec{S}$ that are quite different from the sets X_τ we have considered so far.

Let us start with the tangles of some fixed set S of partitions of V. Consider a tree of tangles $T \subseteq S$ as in Theorem 1. Every tangle τ of S lives at some location σ of T, and we can associate with it the set

$$V_{\tau,T} := \bigcap \{ A \mid A \in \tau \cap \vec{T} \} = \bigcap \{ A \mid A \in \sigma \}$$

of all $v \in V$ that live at that location σ too (see Section 8.2). These sets $V_{\tau,T}$ satisfy the requirement made of hard clusters in Section 6.1 that set them apart from fuzzy clusters: they are subsets of V defined by specifying *for each v individually* to which of them it belongs.

Not every $v \in V$ has to lie in one of the sets $V_{\tau,T}$: those v that live at a location of T that is not home to a tangle of S will lie in none of them. This is good news if we think of interpreting these sets as clusters.

However, the sets $V_{\tau,T}$ need not, in general, be cluster-like at all: they can be sparse, or even empty – for example, when τ is a black hole tangle. As we saw in Section 2.4, however, such black hole tangles can be dissolved by enlarging both S and T: we just have to add enough partitions, nested with T and with each other, that cut through the black hole to 'subdivide it into triangles',[4] as drawn in blue in Figure 14.2. These triangles will no longer be home to a tangle: since they are empty, the triples forming their boundaries are inconsistent when oriented inwards, and thus cannot extend to a tangle of the enlarged S. But neither can the black hole tangle of the original S extend to any tangle of the extended S *other* than the up-closure of a new triangle oriented inwards.[5] The black hole tangle of S has thus been 'dissolved' by the new blue partitions.

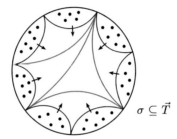

$\sigma \subseteq \vec{T}$

FIGURE 14.2.   Subdividing the black hole into empty triangles.

If we add the blue partitions to T, too, its original location σ will also cease to be a location of T: it will be replaced by the triangles, each oriented inwards. However even the original T can still serve as a tree of tangles for our larger S. Its location σ is simply no longer home to any tangle. Its empty intersection $\bigcap \{\, A \mid A \in \sigma \,\}$, therefore, will no longer feature as a set $V_{\tau,T}$, which is our aim if we want these sets to be cluster-like.

On the other hand, even the void sets $V_{\tau,T}$ carry structural information about the relative position of the other sets $V_{\tau,T}$, those that are more cluster-like. The minimal tree of tangles T that distinguishes the seven cluster-induced tangles of S and the central black hole tangle in Figure 14.2, drawn in black lines, displays a symmetry between the seven 'real' clusters, which would be lost if we subdivided the location σ of the black hole in T into triangles. Indeed, the additional partitions drawn in blue, when added to T, would suggest that those seven clusters were grouped into four super-clusters. However this additional structure would be fake: those seven real clusters are *not* organized into any larger groups.[6] The void $V_{\tau,T}$ of the black hole tangle may not itself be a cluster, but it is a hub around which the clusters are arranged symmetrically. It thus us helps to display their relative structure.

The sets $V_{\tau,T}$ depend not only on the tangles of S but also on our choice of T, the tree of tangles whose existence is ensured by Theorem 1. For example, every non-empty subset of the set S of the three vertical partitions shown in Figure 14.3 could serve as a tree of tangles for the two $\mathcal{F}_6$-tangles of S, one orienting S to the left, and one orienting it to the right. Depending on this choice of $T \subseteq S$, the points in the middle of the central handle could share their location with either of the two tangles, or with neither.

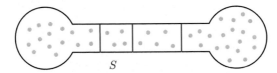

FIGURE 14.3.   Two tangles of S, but two to four locations of T.

In general we will try to make both S and T as large as possible if our aim is to identify point clusters in V associated with our tangles. Figure 14.3 illustrates this well: if S contains enough vertical partitions cutting through the handle to ensure there are no middle tangles,[7] and we then choose T as S, the only clusters we obtain as sets $V_{\tau,T}$ will be the two clusters at either end of the figure – intuitively the ideal outcome.

Let us now consider a tree of tangles $T \subseteq S$ as in Theorem 2. Unlike in our considerations just now, where S was some given set of partitions of V, our setting now is that S is the set of all partitions of V.

For every k, the set $T_k := T \cap S_k$ is a tree of tangles for the k-tangles of S. These T_k, and hence the sets V_τ of points $v \in V$ living at the various locations of the (pre-) tangles of T_k (see Section 8.1), are nested: as k increases, the tree sets T_k get larger, and the point sets V_τ get smaller. Theorem 2 thus provides a hierarchy of these clusters V_τ, in which large but possibly sparser clusters for small k are refined into smaller but denser clusters for larger k.

Figure 14.4 shows four 2-tangles $\rho_1, \ldots, \rho_4$ distinguished by the tree set T_2, whose elements are the partitions of order $k = 1$ shown in green. The leftmost one of these 2-tangles, ρ_1, lives on as a 3-tangle σ_1, because the cluster that induces it is dense enough not to be cut in half by a partition of order $k = 2$. The central 2-tangle, ρ_3, spawns four 3-tangles

$\sigma_2, \ldots \sigma_5$, which are distinguished by the partitions of order $k = 2$ drawn fat in blue, those in $T_3 \setminus T_2$.

The 2-tangles ρ_2 and ρ_4 are criss-crossed by partitions of order $k = 2$ and therefore die at $k = 3$: they do not live on as 3-tangles. The 2-tangle ρ_1, which lived on as 3-tangle σ_1, only dies at $k = 4$ when it is criss-crossed by partitions of order $k = 3$ and therefore does not live on as a 4-tangle. The 3-tangles σ_2 and σ_5 live on as 4-tangles τ_1 and τ_6, respectively.

The central 3-tangle, σ_3, spawns two further 4-tangles, τ_2 and τ_3, which are distinguished by two partitions in T_4 of order $k = 3$ drawn fat in red. Similarly, the 3-tangle σ_4 spans two 4-tangles, τ_4 and τ_5, distinguished by a single partition drawn fat in red.

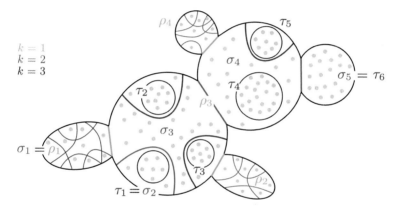

FIGURE 14.4. A tree of tangles T capturing a hierarchy of clusters.

Note that our tree of tangles T is not minimal. Indeed, two of the three central partitions, one blue and two red, would suffice to distinguish the three 4-tangles $\tau_1, \ldots, \tau_3$. We could omit one of those two red partitions from T_4, which would then still distinguish all the 4-tangles. Although the bottom-left blue partition is not needed to distinguish any 4-tangles (if we keep all the red partitions), we cannot delete it from T_4 since we need it in $T_3 \subseteq T_4$ to distinguish σ_2 from σ_3.

The hierarchy of tangles in our example is reflected by a corresponding inverse hierarchy of the clusters $V_{\tau,T}$ they define. Unlike in the example of Figure 14.3, where for $T = S$ the two central locations were not home to a tangle and did not therefore define a cluster of the form $V_{\tau,T}$, the locations of the tree sets T_k in Figure 14.4 are all inhabited by k-tangles of S. When these tangles refine each other as $\tau \subseteq \tau'$, their corresponding clusters refine each other in the opposite way, as $V_{\tau,T_k} \supseteq V_{\tau',T_{k'}}$. For example, we have growing k-tangles $\rho_3 \subseteq \sigma_3 \subseteq \tau_3$ for $k = 2, 3, 4$, which is reflected by shrinking but increasingly dense clusters $V_{\rho_3,T_2} \supseteq V_{\sigma_3,T_3} \supseteq V_{\tau_3,T_4}$.

Unlike with some traditional hierarchical clustering methods, the hierarchy displayed here is not the result of any decisions made by us: the smaller dense clusters are not subsets of the larger loose ones because we chose to partition the latter further, or to partition further the subsets of V of which they were found to be a cluster. Instead, our entire tree of tangles T consists of partitions of the entire set V: the hierarchy of the clusters associated with these tangles is not imposed, but is the result of the substantial fact, offered by Theorem 2, that the partitions in T that distinguish these tangles efficiently are all nested, even across different orders k.

Finally, there is Theorem 3. Given any integer $n \geqslant 1$, it offers a tree set $T \subseteq S$ that witnesses the non-existence of an $\mathcal{F}_n$-tangle of S if indeed none exists. This is, perhaps, the most tangible difference to those traditional clustering algorithms that will always produce some clusters, no matter whether this fits the data or not. Tangle-induced clusters will not be 'found' if they do not exist in the data. Instead, our algorithm from Section 11.4 is designed to find an easily checkable proof of their non-existence.

14.2 Clustering by tangles: fuzzy clusters

Let us now turn to our candidates for 'fuzzy clusters' associated with a tangle from Section 6.1: the subsets of V that guide and witness it.

While most or all of subsets of V that we would intuitively regard as a fuzzy cluster do witness and guide a tangle, the converse is not true. For example, the black hole tangle from Section 2.4, which we discussed in Section 14.1 too, is guided and witnessed by the entire set V. But we would hardly wish to count V as a cluster: there are n well separated clusters in this example, and their union V is not a cluster in any reasonable intuitive sense. Thus, while real clusters, even when fuzzy, do seem to guide and witness a tangle, not all sets that guide and witness a tangle can be regarded as clusters.

This discrepancy is not a consequence of the fact that, in our example, the guiding and witnessing set that resembled no cluster was the entire point set V. For example, if the elements of V are pairwise highly similar with respect to some similarity function that we have good reason to apply – for example, because it has helped us detect clusters in datasets of a similar kind as our current one – we may have detected, rightly, that all of V is one big cluster.

Note that these two examples of tangles, both guided by V, differ in one important respect. Given any similarity function that gives rise to the n clusters in the black hole example, and given the order function it induces via (O1) or (O15), say, the black hole tangle in the centre is likely to be fake in the sense of Section 11.1: it will not extend to any k-tangle of the set S^* of *all* the partitions of V. This is because the partitions that pitch some of the n clusters against the others, such as the blue partitions in Figure 14.2, will all have low order – precisely because they do not cut through any of those n clusters. Hence if k is big enough that $S \subseteq S_k^*$, or just a little bigger, then these additional partitions will be in S_k^* too. Then the n cluster-induced tangles of S will extend to k-tangles of S^*, while its black hole tangle will not.

By contrast, in our earlier example where V was one big cluster because all its pairs were highly similar, this cluster induces a k-tangle of S^* under (O1) or (O15) for all k small enough that only very unbalanced partitions of V have order less than k. And k-tangles of S^* cannot be fake, by definition.

Following these contrasting examples, let us update our attempt to capture tangle-induced 'fuzzy clusters'. So far, we looked for these amongst the sets that guide and witness a given tangle; but the black hole tangles exhibited that this definition is too wide. In any setting where we have an order function on the set S^* of all the partitions of V, let us consider as 'fuzzy clusters' only subsets of V that guide and witness a tangle of S that is not fake: one that extends to a k-tangle of S^* for some k.

We have not arrived at our goal yet. Let us add some more points to the black hole example: a new cluster C of points right in the centre of the black hole (Figure 14.5). The set X of the original points, those that lie in the n clusters around the former black hole, is still a guiding and witnessing set for the original black hole tangle τ, the set of n partitions of $X \cup C$ that split off one of those n clusters and are oriented away from that cluster, now towards C. But this tangle τ is no longer fake now. Indeed, any additional partitions of the new $V = X \cup C$ whose order is no bigger than that of the partitions in τ will have to cut around the new

central cluster C. Hence we can orient them all towards C, to extend τ
to a k-tangle of the set of all partitions of V for $k := \max\{|s| : \vec{s} \in \tau\} + 1$.

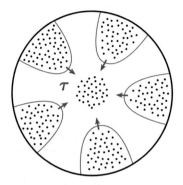

FIGURE 14.5.   The filled black hole: a central cluster C
with $n \geqslant 4$ clusters around it.

The set X in this example, therefore, satisfies our revised attempt
to capture fuzzy clusters, but it is clearly not cluster-like (fuzzy or not).
However, we can refine our definition further.

In the example, an obvious way to improve our non-cluster X would
be to replace it by C, which also guides and witnesses τ but is more
cluster-like than X. We could capture this more generally by requiring that X should guide τ *with maximum reliability*. In the example,
C clearly wins over X in this respect, indeed over all other subsets of V.

This additional requirement removes outliers from guiding sets more
generally, and thus enhances their cluster-likeness. Indeed, as we observed in Section 6.1, any given $v \in V$ will likely lie in *some* guiding
set for *any* tangle that is guided by a subset of V at all, since we could
just add it to such as set X. If v is far from such the tangle's cluster,
however, then adding v to X reduces its reliability as a guide for that
tangle. Hence our revised target sets for clusters, those that guide a
non-fake tangle with maximum reliability, will be more cluster-like than
without the maximality requirement.

However, we now run into a different problem: our target sets for
clusters may become too small. For example, as soon as $\tau = \tau_v$ for some
$v \in V$, the singleton $\{v\}$ guides it with maximum reliability. But if τ has
this form, there is little need to look for a fuzzy cluster in the first place,
since $\{v \in V : \tau = \tau_v\}$ is the perfect hard cluster for τ. Requiring that
τ should not be of this form, therefore, further enhances the similarity
between our target sets and intuitive fuzzy clusters.

Our final attempt, for now, to capture tangle-related fuzzy clusters by some general mathematical conditions is as follows. Let us look for them amongst the subsets of V that *guide a non-principal non-fake tangle with maximum reliability* (amongst all the subsets of V, for the given tangle).[8]

Such subsets of V may well be difficult to find algorithmically; indeed they need not even exist for a given tangle. A dataset with a submodular order function that has a k-tangle not guided by any real-valued function on V, let alone by a subset of V, is described in [19, Example 9]. If the order function is given by (O1), however, even with σ taking arbitrary nonnegative real values, then every tangle in S^* is guided by a function $V \to \mathbb{N}$ [19, Theorem 7]. If the tangle is non-principal and non-fake, we might think of such functions as potentially describing *soft fuzzy clusters.*

Although guiding functions for tangles may not be easy to find, one can characterize those that have them. Details can be found in [10].

14.3 Tangles are not clusters in the feature space

Very informally, tangles differ from clusters as follows. A cluster is a subset of a set of objects, the 'space of points', whose elements are somehow close – for example because, as objects, they share many features. In a set of pieces of furniture, such a cluster might be the subset of chairs. A tangle, by contrast, is a set of features that are shared by many objects. In the furniture example, these features might be having four legs, a surface to sit on, and a straight back. Given this broad picture one may ask, however, whether tangles are simply clusters in the set of those features: clusters in the 'feature space'.

More formally, let $\vec{S}$ be a feature system of partitions of a set V. Every specification of S is a subset of $\vec{S}$, and tangles of S are special such subsets. Are they clusters in $\vec{S}$ with respect to some distance function on $\vec{S}^{(2)}$? And if so, are they *nothing but* clusters in $\vec{S}$: is every cluster in $\vec{S}$ under this distance function a tangle?

If the answer to these question was positive, finding tangles would be a new type of clustering in $\vec{S}$: no less, but also no more. We would then have to compare the tangles in $\vec{S}$ with its more traditional clusters. Conversely, we might ask how traditional clusters in V compare with tangles in $\vec{V}$, the feature system dual to $\vec{S}$ in the sense of Section 7.7.

However, tangles are not merely a new kind of 'cluster in the feature space'. They differ from those in two crucial aspects.

The first is that the elements of $\vec{S}$ which a tangle selects are not potentially arbitrary subsets of $\vec{S}$: they all have to pick exactly one of the two elements $\vec{s}$, $\overleftarrow{s}$ associated with every $s \in S$. In particular, tangles – viewed as subsets of $\vec{S}$ – will never be disjoint: different tangles will have large subsets of $\vec{S}$ in common.

Figure 14.1 illustrates this well: it has two k-tangles for $k > \ell$, but as subsets of $\vec{S}$ these differ only by a single element: while one contains $\vec{t}$, the other contains $\overleftarrow{t}$. For all other $s \in S_k$, both tangles contain the specification $\vec{s}$ of s that orients it inwards. Thus, while these two tangles in $\vec{S}$ capture well the fact that the set V in this example has two clusters – which each guide and witness one of the two tangles – the set $\vec{S}$ has only one cluster induced by V, by any clustering standards.

This difference is somewhat mitigated in applications, such as those in Sections 6.3 or 6.4, where the elements s of S come with a default specification $\vec{s}$ that is deemed more important than its inverse. In this case we might ignore in a tangle τ those s which it specifies as $\overleftarrow{s}$, and think of it as the only the subset $\{\, \vec{s} \in \vec{S} \mid \tau(s) = \vec{s} \,\}$ of $\vec{S}$. But these subsets are still likely to intersect substantially between different tangles, unlike distinct clusters: the tangle of 'green' items in a shop, say, is likely to include many pricier-than-average items, and vice versa.

Another fundamental difference between tangles and clusters in $\vec{S}$ is that clusters are usually defined in terms of pairs of objects (here: features), whereas tangles are crucially defined in terms of triples. See Chaper 7.3 for why this difference is so important, both for the notion of a tangle and for the theory built on this notion, which we seek to apply now.

The pie example in Figure 7.3 illustrates this latter difference between tangles and clusters in $\vec{S}$. Each of the 2^n specifications of S in that example is a pre-tangle. This makes it into a 'cluster' in $\vec{S}$ in the sense that its elements are pairwise close. (Recall that $\vec{r} \cap \vec{s}$ is large for every pair of orientations $\vec{r}$, $\vec{s}$ of distinct $r, s \in S$, because it contains a cluster in V; so $\vec{r}$ and $\vec{s}$ are 'close' elements of $\vec{S}$.) By contrast, the pie example has only $2n$ tangles. But these correspond to the $2n$ visible clusters in V.

14.4 Tangles versus decision trees and hierarchical search

Consider a questionnaire Q, consisting of yes/no questions as usual. The various ways of answering Q can be depicted by a graph-theoretical tree, as follows. We enumerate the questions in Q, say as $q_1, \ldots, q_m$, and start our tree by drawing a single node, the *root*. We then draw the tree level by level, starting from the root as level 0, by adding the ith level for $i = 1, \ldots m$ as follows. To every node at level $i - 1$ we attach two new nodes as its *children*, one for a yes-answer to question q_i and another for a no-answer. The ith level of our tree thus has twice as many nodes as the $(i - 1)$th level; all in all, our tree will have $\sum_{i=0}^{m} 2^i = 2^{m+1} - 1$ nodes.

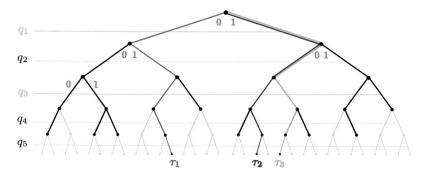

FIGURE 14.6. The decision tree for five questions. Three of the 32 specifications of Q form tangles for Figure 14.7.

Figure 14.6 shows such a binary tree for $Q = \{q_1, \ldots, q_5\}$.[9] Every specification of the entire set Q of questions marks a path in this tree, from the root at the top all the way down to a *leaf*: a sequence of yes/no answers to the questions $q_1, \ldots, q_m$.

If our questionnaire has been answered by a set V of people, then every person $v \in V$ specifies Q by their set $v(Q)$ of answers. This marks a leaf of our tree, and we can associate v with this leaf. If some particular set of answers was chosen by at least n people, then this set of answers is an $\mathcal{F}_n$-tangle of Q. In Figure 14.7, this is the case for three sets of answers, marked τ_1, τ_2, τ_3, for $n = 8$ (say).

These particular specifications of Q, indicated in Figure 14.6 as $\tau_1 = (0, 1, 1, 0, 1)$ and $\tau_2 = (1, 0, 0, 1, 0)$ and $\tau_3 = (1, 0, 1, 0, 0)$, can thus be identified as tangles simply because enough elements of V, at least n, agreed with them on every question in Q. In our usual terminology, these are the $v \in V$ such that $\tau_v \in \{\tau_1, \tau_2, \tau_3\}$. In particular, the tangles τ_1, τ_2, τ_3 in our figure are all principal. Principal tangles τ_v of Q are associated with the same leaf of the decision tree for Q as their defining v is, and can thus be readily identified from this tree.

Most tangles, however, are non-principal, including tangles guided and witnessed by a cluster in V. If τ is a non-principal tangle of Q,

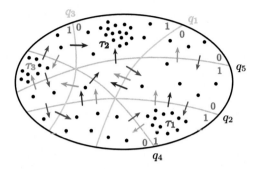

FIGURE 14.7.   Tangles τ_1, τ_2, τ_3 of Q, with tree of tangles $\{q_1, q_3\}$.

there is no $v \in V$ that agrees with τ on every question. If we view the elements $\vec{q}$ of τ as subsets of V, as usual, this is expressed by $\bigcap \tau = \emptyset$. Such tangles τ cannot readily be identified from the decision tree corresponding to Q, since no $v \in V$ is associated with the leaf marked by τ. To verify that τ is indeed a tangle, we have to check for every triple of questions whether at least n people agree with τ on *those* three questions.

Summing up, the decision tree for a questionnaire Q illustrates the set of all specifications of Q, not just the set of its tangles. Whether or not a questionnaire answered by V has a tangle at all is a non-trivial question, whose answer can tell us a lot about the structure of V. But every questionnaire has a decision tree. Unlike tangles, this tells us nothing about any structure among the people that have answered the questions.

This difference is illustrated well by our example of expert systems, in particular for medical diagnosis, in Section 4.1. The simplest kind of expert system will ask certain questions $q_1, \ldots, q_m$ designed to narrow down the cause of an event to be explained. In medical diagnosis, these might check a list of potential symptoms or consist of simple measurements. The entire process is known as 'excluding' known illnesses as the cause of a patient being sick: those illnesses that would give a different reading on at least one question q_i from the reading taken on the patient. Ideally, the set of answers, a complete specification of Q, will narrow the options down to just one illness as a possible cause, which is then thought to have been established as the true cause. However, in reality things are not as mechanical as this: readings can be faulty or influenced by other factors. This is why tangles offer such a new and different approach to diagnosis in 'fuzzy' environments. The difference between tangles and decision trees illustrates this: while all sets of readings mark equally valid paths in the decision tree, those that are tangles correspond to just a few of those paths that are more typical than others.

The picture is simplified somewhat if we replace Q with a tree of its tangles, T say, as in Theorem 1. The partitions of V in T are nested, and all their pre-tangles are tangles of T. If T was chosen minimal, then each of its locations is inhabited by a tangle – even one induced by a tangle of Q. These tangles of T mark leaves of the decision tree for $T \subseteq Q$,[10] but these are not all its leaves: every inconsistent specification of T also marks a leaf of its decision tree. The leaves marked by a tangle of T are associated with those $v \in V$ that live at its location in T. But there may be no such v for a given tangle: recall that even empty locations of a tree set may be inhabited by a tangle.[11] In Sections 8.2 and 14.1 we denoted the set of those v as $V_{\tau,T}$, and considered these sets as candidates for clusters in V.

The fact that the tangles τ of Q and the sets $V_{\tau,T}$ are associated with leaves of the decision tree for T is a far cry, however, from saying that tangles are no more than paths in a decision tree: unlike with the decision tree for Q, which we could simply write down (even independently of V), the tree of tangles T is the result of our tangle analysis of Q in V. The fact that the tangles of Q can be illustrated by the decision tree for T does not distinguish them from the other specifications of T, the inconsistent ones, and thus has no intrinsic value.

Decision trees do, of course, illustrate our process of *finding* the tangles of a given feature system $\vec{S}$, as described in Section 11.2. Their branchings at the level corresponding to an element s of S correspond to its two possible specifications, $\vec{s}$ and $\overleftarrow{s}$. Every path down from the root maps out a specification of all of S, as earlier for Q. But rather than continuing all these paths down to a complete specification of S, we do not branch along features $\vec{s}$ that would create an inconsistency with the earlier features on its path.

In particular, we stop extending a partial specification σ of S marked by the path from the root down to our current node whenever neither of the features $\vec{s}$ and $\overleftarrow{s}$ branching out of that node can be added to σ without creating an element of $\mathcal{F}$ in it. Whenever this happens, σ is a maximal tangle in $\vec{S}$, a k-tangle for $k = |s|$. The partial binary tree obtained in this way illustrates how we constructed the tangles in $\vec{S}$, all of them. In Figure 14.6, this is the tree shown in colour and black.

Finally, decision trees also formalize what is known as *hierarchical clustering*: the process of iteratively partitioning V into smaller and smaller subsets. The decision tree that visualizes this process has these subsets as nodes, and when a subset is split in two then these parts become the children of the node representing the earlier subset. The best way to split a given subset A in two is worked out only once A has been computed, so this method is less crude than splitting the subsets obtained along pre-given partitions of V, as we did earlier when these splits were given by a fixed questionnaire. In hierarchical clustering we would, in that case, be allowed to come up with a new question to put to the population in a subset we had singled out by our earlier question, one that might be particularly relevant to these people and less to others.

The subsets obtained by such an iterative splitting process are clearly nested: they form a tree set of subsets of V, much like the sets $V_{\tau,T}$ living at the various locations of a tree of tangles T. The difference is that our T consists of partitions carefully chosen to cut V between potential clusters inducing tangles, which were computed taking all the potential partitions of V into account. In practice, these will be partitions of low order, the 'bottleneck' partitions of V. A hierarchical clustering algorithm will also seek to split each given subset at its 'bottlenecks'. Their bottleneck quality, however, would be measured and decided in the subset to be split, rather than in all of V as in the case of tangles.

It is difficult to say in general which approach is better. However it is also possible to combine them. Suppose our hierarchical clustering algorithm tells us to split V into A and A', and then A into B and C, and A' into B' and C'. We might then add to our system $\vec{S}$ whose tangles we aim to compute not just these six subsets A, B, C and A', B', C', but also the sets $B \cup B'$ and $B \cup C'$ (as well as their complements $C \cup C'$ and $C \cup B'$). Phrased more globally, we would use the hierarchical clustering method to split V recursively into disjoint 'atomic' subsets whose overall union is V, but would recombine these in all possible ways into partitions of V into just two subsets, and then compute the tangles of those partitions of V. This approach has been used, amongst other methods, in [3].

14.5 Clustering by advanced feature tangles, versus spectral clustering

Tangles of advanced feature systems can be used directly to define hard clusters of V, without a need to convert those features into partitions of V first. We shall illustrate this in this section for the case of our spectral feature systems from Section 10.2. But it applies equally to the PCA-based advanced feature systems we found in Section 10.3.

Our spectral construction of an advanced feature systems begins just like classical spectral clustering performed on V: by computing a sequence $f_1, \ldots, f_\ell$ of normed orthogonal eigenvectors f_i for the first ℓ eigenvalues $\lambda_1 \leqslant \ldots \leqslant \lambda_\ell$ of the Laplacian L associated with a similarity measure $\sigma \colon V^{(2)} \to \mathbb{R}$ of our choice, where $\ell \leqslant n = |V|$ is a parameter we can choose, and σ is assumed to take non-negative values only.[12]

Standard spectral clustering proceeds from here by embedding V in $\mathbb{R}^\ell$ by mapping each $v \in V$ to the ℓ-tuple $(f_1(v), \ldots, f_\ell(v))$, then uses some standard clustering algorithm, such as k-means, on this representation of V in $\mathbb{R}^\ell$, and finally translates the clusters found back to V.

Our Section 10.2 approach instead puts the eigenvectors $f_1, \ldots, f_\ell$ in $\vec{S}$, the feature system of which we then computed our tangles – possibly after adding some infima and suprema to make it more submodular. Suppose we have done this now, and for simplicity assume that we computed only the tangles of S itself, not of a hierarchy $S_1 \subseteq S_2 \subseteq \ldots$ in S. Each of our tangles τ then specifies each of the features $f_1, \ldots, f_\ell$, either as $-f_i$ or as f_i, so we can view τ as a map $\{f_1, \ldots, f_\ell\} \to \{-1, 1\}$.

In our earlier approaches in this chapter to using tangles for point clustering in V, the only information about the points $v \in V$ we used when forming those clusters was how they specified $\vec{S}$. When $\vec{S}$ is built from $f_1, \ldots, f_\ell$, we would thus remember for v precisely the information $(f_1(v), \ldots, f_\ell(v))$.[13]

This is exactly what standard spectral clustering remembers about V too. The difference is how we interpret this information about the points in V, and then what we do with it. For example, if $f_1, \ldots, f_\ell$ were used to define partitions of V, we remember for every $v \in V$ where it lies relative to each of these partitions; but we keep thinking of it as a point in V. In the spectral clustering world, we essentially forget V and replace it with $\mathbb{R}^\ell$, in which every $v \in V$ has its place via $(f_1(v), \ldots, f_\ell(v))$.

Spectral clustering then clusters these points in $\mathbb{R}^\ell$ by some standard method, typically based on their distance in $\mathbb{R}^\ell$. By contrast, we can measure directly how well a tangle τ of $\{f_1, \ldots, f_\ell\}$ represents a point v: by the dot-product of the ℓ-tuples $(v(f_1), \ldots, v(f_\ell))$, where $v(f_i) = f_i(v)$ as usual, and $(\tau(f_1), \ldots, \tau(f_\ell))$, where $\tau(f_i) \in \{-1, 1\}$ as earlier. Our clustering is not done in $\mathbb{R}^\ell$, let alone by comparing pairs of points in $\mathbb{R}^\ell$: it was essentially done when those τ were determined. What remains is only to associate each v with the tangle that represents it best.

14.6 Image segmentation, identification, compression

In Section 6.2 we saw how, in principle, tangles can be used to capture visible 'regions' of an image. Essentially, these would be modelled as clusters in the *canvas*, a set V of pixels, based on a similarity function that takes account both of the proximity of pixels and their similarity in appearance, such as their colour, brightness and so on. In order to use tangles to capture such regions, we first have to agree on a set S of partitions of V. We already discussed that we might limit these to *cuts*, partitions by a single line cutting the canvas in half.

Let us now revisit this question, of how to choose S, in the light of the tangle theory we have met since Section 6.2.

For a start, we shall wish to add to S at least some corners of partitions already in S, so let us do that right away. In practice it will usually suffice to do it once, or perhaps take corners of corners one more time. But the resulting set S of partitions whose tangles we compute will still be sufficiently restricted. In order to simplify our discussion we shall therefore assume, unless otherwise stated, that the partitions we are talking about are cuts: paritions with a single dividing line.

Next, we have to choose an order function. This will not have to be generic, as discussed in Chapter 9, but can appeal to the particular kind of partitions whose order we have to define. Although we did not have the notion of an order function available in Section 6.2, we already discussed the basics of how to choose one: the order of a cut of the canvas will be something like the average of the similarities $\sigma(p, q)$ of pairs of pixels that are adjacent and divided by the cut. The adjacency here will probably have to be relaxed a little, so that pairs (p, q) are also taken into account when they are divided by the cut but are not adjacent, just close. Very roughly, the closer they are, the more weight shall we give to their similarity $\sigma(p, q)$.

With such an order function in place, we can look at the hierarchy of the k-tangles in S as k increases, as discussed in Section 7.4. All our tangles shall have agreement at least 2, probably a little bigger, so that there are no tangles focussed on a single pixel. Let us now see how our order function is reflected in the regions captured by those tangles.

Figure 14.8 shows a simple canvas consisting of black and white pixels only, the shape of the letter L in black on a white background. Let $\sigma(p, q)$ be 20 if p and q are both black, 10 if they are both white, and 0 if one is white and the other black. Given a cut s of the canvas

FIGURE 14.8.   A hierarchy of regions from a hierarchy of tangles.

defined by a single line, let its order be

$$|s| := \sum \sigma(p,q)/\ell(s)$$

where the sum is taken over all 2-sets $\{p,q\} \subseteq V$ such that the pixels p and q are adjacent on the canvas but lie on different sides of s (so that their shared pixel boundary lies on the dividing line of s) and $\ell(s)$ is the number of such 2-sets.

For $k = 1$, there is a single partition of V in S_k: the partition that cuts out the entire shape L by a single line around it. Note that $\sigma(p,q) = 0$ for every pair of pixels p,q evaluated in $|s|$, so $|s| = 0 < k$. This partition s has two orientations, one towards the L and the other away from it. Each of these forms a tangle of order 1, so we have two 1-tangles in S. They correspond to the two most obvious regions: the L-shape and its background.

Now let k be a little bigger, perhaps $k = 2$: just big enough that the cut by the green line lies in S_k. There are many similar such lines, all consisting of a short vertical segment cutting through the horizontal shaft of the L, followed by a line round the L along its boundary. The order of these cuts grows slightly as the vertical line moves to the left, because it makes up a larger proportion of the entire green line then. But once they are all in S_k, any k-tangle orients them all inwards or all outwards. (Compare what we noted in Section 2.4 about the cuts through the handle in Figure 1.3.) Together with the unique cut of order 0, these cuts have three k-tangles: one that corresponds to the white background, one that corresponds to the region consisting of the fat vertical shaft of the L, and a third consisting of the serif on the right.

This last tangle orients the cut around the L-shape inwards and all the green cuts outwards. Note that it is not necessary, for this tangle to exist, that the partitions which cut vertically through the bottom shaft and then follow the L-shape round the serif on the right are also in S_k: they will have higher order than the green ones, since the proportion of their cutting line where σ is positive is greater. Once k is just large enough that these cuts are in S_k too, perhaps for $k = 3$, there will be more partitions in S_k but still only the 'same' three tangles as before: those corresponding to the background, or to the vertical shaft, or to the serif on the right.

Now let k become big enough, perhaps $k = 4$, that the cuts in S_k include all those whose dividing line cuts through the serif on the right, as the yellow line does, and then follow the black shape round either way. These cuts will destroy our earlier tangle that corresponded to the serif on the right, because we excluded all focussed tangles; compare our discussion in Section 7.4 around Figure 7.4. So for this k we only have two tangles left: one corresponding to the white region, and another corresponding to the black vertical shaft.

When k is finally large enough that cuts through the vertical shaft (as indicated by the red line) which then follow the L-shape round are in S_k, too, there will be only one tangle left: that corresponding to the white background.

Let us now see how tangles of cuts of an image can be related to regions in that image more generally. The idea is to think of a smaller and a bigger tangle as 'the same' if the bigger one induces the smaller and every tangle of intermediate order that does so too is in turn induced by the bigger one. Formally, if S is a set of cuts of V of varying order, we call a k-tangle τ and a k'-tangle τ' in $\vec{S}$ with $k \leqslant k'$ *equivalent* if every ℓ-tangle ρ with $k \leqslant \ell \leqslant k'$ that satisfies $\ell \cap \vec{S_k} = \tau$ also satisfies $\tau' \cap \vec{S_\ell} = \rho$.

Note that equivalent tangles are indistinguishable, because the larger of the two induces the smaller. In particular, tangles of the same order cannot be equivalent unless they are equal. Conversely, indistinguishable tangles are not normally equivalent. Indeed, for $k < k'$ consider a k-tangle τ that is induced by two k'-tangles τ' and τ''. Then τ is indistinguishable from both τ' and τ'', but equivalent to neither, since τ' and τ'' can each play the role of ρ from the definition of equivalence and neither induces the other.[14]

We noted that equivalent tangles of the same order are identical. Equivalent tangles of different orders $k < k'$ are also essentially the same: the bigger one is the *unique* k'-tangle that induces the smaller k-tangle, which conversely is not induced by any ℓ-tangle with $\ell \leqslant k'$

other than (that induced by) the bigger one, not even an ℓ-tangle that does not extend to any k'-tangle. In the terminology of the hierarchy of tangles outlined in Section 7.4, a smaller tangle 'grows into' an equivalent bigger one – rather than dying on the way to spawn several new tangles, of which the bigger tangle is only one.

In our application of tangles to image analysis, equivalent tangles describe essentially the same region of the image. The larger of two equivalent tangles just cuts around this region more tightly, because its partitions are allowed to have larger order and can thus encroach on the region more closely, cutting away outgrowths that the lower-order cuts in the smaller tangle cannot cut away.

Let us then use tangle equivalences to formalize the intuitive notion of a 'region' in an image, and of how regions are related. Note that this formalization does not refer to the elements of the canvas V, just to tangles. In particular, just as we strove in Section 14.1 to describe generic clusters in terms of tangles but not as subsets of V, we can now describe regions in terms of tangles rather than as subsets of V.[15]

Given S, a *region* of V is an equivalence class of tangles in $\vec{S}$. As soon as a tangle τ' from a region R' induces a tangle τ from a region $R \neq R'$, every tangle in R' induces every tangle in R. We then say that R' *lies inside* R. This can indeed happen since, as we have seen, τ' can induce τ without being equivalent to it. A partition in S *distinguishes* two regions if it distinguishes two tangles representing those regions. Just as indistinguishable tangles need not be equivalent, distinct regions can be indistinguishable;[16] this is the case if and only if one lies in the other.

The *complexity* of a region is the smallest k for which it contains a k-tangle; its *cohesion* is the largest such k. The numerical difference between the two, a region's cohesion minus its complexity, is its *visibility*.

To see what it means for a region R to have complexity c, consider any $k < c$. Pick a tangle σ that represents R, and let τ be the k-tangle it induces. Note that τ depends only on R, not on the choice of σ. Since σ induces τ but is not equivalent to it, by $k < c$, we know that τ is also induced by some tangle ρ of order $\ell \leqslant c$ that is not induced by σ. In other words, ρ is not the ℓ-tangle that σ induces. So ρ is not equivalent to τ either. Thus, τ is induced by two tangles, σ and ρ, that are both not equivalent to it, and of which neither induces the other. Hence the regions they represent, R and R', say, are distinct regions that both lie inside the region that τ represents. For $k = c - 1$ this means that τ is a $(c-1)$-tangle which spawns at least two c-tangles, those induced by σ and ρ. The complexity of R is the point in time, in the evolution of tangles as described in Section 7.4, at which it arises as a smaller subregion inside a larger region, the region which τ represents.

Let us now see what it means for R to have cohesion d. One possibility is that there is no k-tangle for any $k > d$ that induces a tangle representing R. Then R has no further regions inside it. In fact, it can be shredded into singletons $\{v\}$ by partitions of order no greater than d; note that these do not guide any unfocussed tangles, those we are considering. The other possibility is that there are several tangles of order $k > d$ that induce the tangles in R, but these induce at least two different $(d+1)$-tangles. The regions that those $(d+1)$-tangles represent lie inside R. In terms of tangle evolution, the first of the above possibilities means that the unique tangle of order d in R dies. The second possibility means that it spawns at least two $(d+1)$-tangles.

In order to visualize the visibility of regions, it may help to look at a concrete example. Figure 14.9 shows a canvas which, intuitively, has two greenish regions and two reddish ones, which combine into a larger 'green' region and a larger 'red' region. The two red squares (one pink, one scarlet) are harder to distinguish than the two green squares (one jade, one lime): unlike those, they are less visible as individual regions than as a combined region. In order to determine its regions formally, let us compute the tangles for this canvas: its k-tangles as k grows.

FIGURE 14.9.   Regions of different visibility.

Let $k_1 - 1$ be the order of the vertical partition s_1 down the middle of the canvas. This k_1 is the smallest k for which we have more than one k-tangle: a tangle τ_1 that orients s_1 to the left, and a tangle ρ_1 that orients it to the right.

Let k_2 the smallest k for which we have more than two k-tangles. Then $k_2 - 1$ will be the order of the partition s_2 of the canvas into the lime green square versus the other three squares. Note that $k_2 > k_1$, because where the dividing lines for s_1 and s_2 differ, the pixels across s_2 are from different greenish squares, and hence more alike, than the pixels across s_1 between the jade square and the pink one. For $k = k_2$ we have exactly three k-tangles. One is the tangle ρ_2 that orients both s_1 and s_2 towards the red half of the canvas and thus extends ρ_1. By contrast, τ_1 spawns two k_2-tangles: a tangle τ_2 that orients s_1 and s_2 towards the lime green square, and a tangle τ_2' that orients them both towards the jade square. The jade square's partition s_3 against the other three squares has higher order than $|s_2|$, since its pixels differ from the pink ones less than the lime green pixels differ from the scarlet ones. But the jade green square is still where one of the three k_2-tangles is located, even though this tangle does not specify s_3 explicitly.

However, $|s_3| > |s_2|$ is still smaller than the orders of both the partition s_4 between the pink square and the rest, and the horizontal partition s_5 along the middle. This is because the latter two partitions have pink versus scarlet pixels along their dividing lines, and these pixels are very similar. Thus for $k = k_3 := |s_3| + 1$ we still only have three k-tangles: the three k_2-tangles ρ_2, τ_2 and τ_2' each extend to a unique k_3-tangle, ρ_3, τ_3 and τ_3' say, because in each case only one orientation of s_3 is consistent with their other elements.

Let $k_4 := |s_4| + 1$. While τ_3 and τ_3' extend uniquely to k_4-tangles, τ_4 and τ_4', say, ρ_3 spawns two k_4-tangles: a tangle ρ_4 orienting s_4 towards the pink square, and another tangle ρ_4' orienting it away from it. We thus have four k_4-tangles now, one pointing to each of the four squares. Note that s_5, the horizontal partition along the middle, has an order too large to be oriented by these tangles. But when k is finally large enough to orient s_5, too, all four tangles extend to unique k-tangles, so no new tangles are spawned.

Our definitions of regions and their complexity, coherence and visibility, capture our intuitive notion of these as follows. The tangles τ_1 and ρ_1 are those of lowest complexity: they are born first, as k_1-tangles. While τ_1 spawns distinct k_2-tangles τ_2 and τ_2', the tangle ρ_1 extends to the unique k_2-tangle ρ_2. This is equivalent to ρ_1, because no other k-tangle with $k_1 \leqslant k \leqslant k_2$ extends ρ_1. By contrast, neither τ_2 nor τ_2' is equivalent to τ_1, because the other of the two extends τ_1 too. However, they are equivalent to τ_3 and τ_3', respectively, while ρ_2 is equivalent to ρ_3.

To see that ρ_1 is also equivalent to ρ_3, without using transitivity,[17] notice that k_2 is the only k with $k_1 < k < k_3$. Since ρ_2 is the only k_2-

tangle extending ρ_1, and is itself induced by ρ_3, the tangles ρ_1 and ρ_3 are equivalent.

Finally, τ_3 and τ_3' are equivalent to τ_4 and τ_4', respectively, and hence so are τ_2 and τ_2'. But ρ_3 is equivalent to neither ρ_4 nor ρ_4', since these are distinct k_4-tangles both extending it.

These equivalences mean that we have the six regions we would intuitively expect: a 'red' vertical region $R = \{\rho_1, \rho_2, \rho_3\}$ in which lie the pink and the scarlet square regions, $R' = \{\rho_4\}$ and $R'' = \{\rho_4'\}$ say, and a 'green' vertical region $G = \{\tau_1\}$ in which lie the lime green square region $G' = \{\tau_2, \tau_3, \tau_4\}$ and jade square region $G'' = \{\tau_2', \tau_3', \tau_4'\}$. The regions R and G have the lowest complexity, k_1. While G has cohesion only $k_2 - 1$, and hence relatively low visibility, the region R has cohesion $k_4 - 1$ and hence much bigger visibility. This is in line with our feeling that the right half of the canvas is more cohesive, and hence more visible as a region, than the left half which divides too readily into its two green squares.

The green squares, however, are intuitively more clearly visible as separate regions than the two red squares are, which are harder to distinguish. And indeed, the green squares have higher visibility than the red ones. To see this, note that all our four square regions have the same cohesion: they each contain k-tangles up to some maximum value k_∞, above which k will be big enough that partitions of order less than k can have dividing lines whose adjacent pixels across the line all have the same colour. Any such k-tangles for $k > k_\infty$ will be focussed; see Section 7.4. If $\mathcal{F}$ contains the singletons $\{\{p\}\}$, as we assumed, there will be no k-tangles for $k > k_\infty$. Either way, the cohesion of each of the square regions will be k_∞.

Hence the green square regions (which, despite their low complexity of k_2, do not split into smaller regions lying in them) will have visibility as high as $k_\infty - k_2$. The two red squares, whose complexity k_4 is much higher, have visibility only $k_\infty - k_4$. This, too, agrees with our intuitive notion of visibility.

Let us conclude this example by noting how the match between our formal notion of the regions of an image, including their complexity, cohesion and visibility, and our intuitive notions of these is affected by how well our set S of partitions captures the intuitive 'bottlenecks' in our image, its natural cutting lines. Our example suggests that our formal notions capture those intuitive ones well – but our discussion rested on the assumption that S consisted of all the cuts of the image, not just some. In practice, this will rarely be the case, since there are too many cuts to process.

If S contains only a selection of all cuts, it is important that the most natural cuts, such as those down the middle of the canvas or around the small squares in Figure 14.9, are contained in S. But it is also important that S contains enough of the not-so-natural cuts, those through one of the squares, to get the cohesion, and hence the visibility, of those regions right. If the jade square has cohesion d, say, then for $k > d$ we can cut it up into singletons by lots of partitions of order at most k. But unless enough of those partitions are in S, our algorithms will not notice that, by zooming all the way in to tangles focussed on single jade pixels. Instead, they will still find k-tangles guided by some subsets of the jade square, and those subsets may appear as 'regions' inside it that do not correspond to intuitive regions.

Next, let us study what our tangle theorems from Chapter 8 can contribute to image analysis. The tree-of-tangles theorem, Theorems 1 and 2, provide us with at tree set $T \subseteq S$ that distinguishes the tangles in $\vec{S}$. Tree sets of cuts across a canvas are easy to describe: they are sets of cuts whose dividing lines do not cross. Our tree set T contains only the most important such lines, those that distinguish different tangles. In our application, these are cuts between different regions.

In the case of Theorem 2, where T distinguishes tangles of different orders efficiently, a region R may lie inside another region Q. This means that a tangle σ representing R refines a tangle τ representing Q that is also refined by a tangle σ' not equivalent to σ. The region R' which σ' represents then also lies inside Q, but not inside R. If R and R' are the only regions inside Q, the tree set T from Theorem 2 will contain either a cut around Q and one around R or R', or cuts around R and R' but none around Q. Hence only two of these three regions will be outlined by lines stemming from cuts in T.

Our tree of tangles T not only provides us with non-crossing lines between different regions, it also helps us determine something like a boundary of those regions. These come from the locations in T of the tangles τ in $\vec{S}$ that represent the various regions. Recall from Section 8.1 that these locations are stars of partitions, in our case stars of cuts. The dividing lines of these cuts, then, mark something like a boundary of the region represented by any tangle that inhabits this location. The region itself is marked by the subset $V_{\tau,T}$ of the canvas that is the intersection of the elements of this star, as seen in Section 14.1.

Ideally, the dividing lines of those few cuts in T are something like a caricature: just a few lines marking the most distinctive divisions between the natural regions of the picture. Since the elements of T distinguish the tangles in $\vec{S}$ efficiently, they mark the clearest possible lines between different regions, just what we expect of a good caricature.

Caricatures are the epitome of what, in the digital age, we have come to call compression: they capture the essence of the object depicted by just a few strokes. If our tree of tangles for an image is canonical,[18] this tree set and its locations also capture something like the essence of that image. Rather than comparing two images pixel by pixel, we might compare their trees of tangles first: if these are substantially different, we can conclude that the two images are not the same, or even that they do not show similar things such as people or landscapes. Conversely, if the trees of tangles of two images contain structurally similar subtrees we might conclude, or at least suspect, that they depict similar things.

When we represent the regions of our image by the sets $V_{\tau,T} \subseteq V$ that correspond to the locations of T which accommodate the tangles τ representing these regions, we can compress the image by representing each of those regions by just a few pixels from $V_{\tau,T}$.

Representations of an image by tangles, no matter how we do this in detail, are robust against small changes in the data, because the tangles themselves are. This can help with communicating the essence of images electronically, e.g. by sending their tangles through a 'noisy channel'.

Finally, there is Theorem 3, which gives us a tree set $T \subseteq S$ that witnesses the non-existence of tangles of a given type, i.e., of $\mathcal{F}$-tangles for a given $\mathcal{F}$. This can be applied directly to give a proof that an image does *not* contain reliable information, e.g. because it is blurred. More precisely, the complexity and the cohesion of a region give us an indication of the minimum or maximum 'resolution' of our camera at which this region can be identified. We thus obtain a mathematically rigorous definition of the adequacy of a particular 'camera resolution' for a given pixellated image.

14.7 E-Commerce: customer-product duality

Let us rephrase our example from Section 6.3 of an online shop, and the tangles of the dual feature systems this gives rise to, in terms of set partitions as introduced in Section 7.2.

In the example, V was a set of customers of an online shop, and S the set of items sold at this shop. We assumed that each customer v has made a single visit to the shop, specifying $s \in S$ as $v(s) = \vec{s}$ if they included s in their purchase, and as $v(s) = \overleftarrow{s}$ if not.[19] In the dual setup, every item $s \in S$ specifies every $v \in V$: as $s(v) = \vec{v}$ if v bought s, and as $\overleftarrow{v}$ otherwise. Thus, $v(s) = \vec{s}$ and $s(v) = \vec{v}$ mean the same thing: that v bought s.

In the terminology of Section 7.2, every item $s \in S$ is a partition of V: the partition $s = \{\vec{s}, \overleftarrow{s}\}$ of V into the complementary sets

$$\vec{s} = \{v \in V \mid v \text{ bought } s\}$$

of customers that bought s and the set $\overleftarrow{s} = V \smallsetminus \vec{s}$ of those that did not. Similarly, every customer $v \in V$ is a partition of S: the partition $v = \{\vec{v}, \overleftarrow{v}\}$ of S into the complementary set

$$\vec{v} = \{s \in S \mid v \text{ bought } s\}$$

of items that v bought and the set $\overleftarrow{v} = S \smallsetminus \vec{v}$ of items that v did not buy.

We thought of $\vec{v}$, or equivalently of the entire specification $v(S)$ of S, which $\vec{v}$ determines, as the 'shopping basket' of customer v. Similarly we thought of $\vec{s}$, or equivalently of $s(V)$, as the 'popularity footprint' of the item s in V.

The tangles of S, specifications of S 'typical for V', were hypothetical shopping baskets reflecting the purchasing behaviour of many customers. If we mentally identify a customer v with their purchase, tangles of S can be thought of as types of customer. Similarly, tangles of V were typical popularity footprints, or types of item: ways of splitting the set of customers into fans and non-fans which broadly reflect the popularity footprint of many items.

More formally, a tangle τ of S of agreement at least n is a specification of S such that for every three elements s_1, s_2, s_3 of S there are at least n customers v that bought or failed to buy each s_i as prescribed by τ: such that $v(s_i) = \tau(s_i)$ for all i, or more explicitly, such that v bought s_i if $\tau(s_i) = \vec{s_i}$ and v did not buy s_i if $\tau(s_i) = \overleftarrow{s_i}$. Similarly, a tangle ρ of V of agreement at least n is a specification of V such that for every three customers v_1, v_2, v_3 in V there are at least n items s that were bought or not by each v_i as prescribed by ρ: such that $s(v_i) = \rho(v_i)$ for all i, or more explicitly, such that v_i bought s if $\rho(v_i) = \vec{v_i}$ and v_i did not buy s if $\rho(v_i) = \overleftarrow{v_i}$.

In practice, what bites in these conditions are only the forward-oriented entries of those triples. This is because an average single purchase contains only a fraction of the items on offer. For example, once we have found a few more than n customers that bought s_1 and s_2, we shall probably find n among them who failed to buy s_3, but we may not be able to find n among them who also bought s_3. Similarly, if customers v_1 and v_2 bought a few more than n common items, it will be easy to find among these n items that customer v_3 did not buy, but there may not be n among these items bought by v_3 too.

But the fact that the partitions v of S, and the partitions s of V, are so unbalanced has more serious implications. For example, if an average purchase consists of 10 items, say, but there are a thousand items in the shop, then each item is included in only about one in a hundred purchases on average, and each triple of items in only about one in a million purchases. Then, for tangles of S, any agreement of $n > |V|/1.000.000$ would count as 'big', and we may need a million customers or so before any tangle analysis of their purchases can be fruitful.[20]

In situations such as this, when the natural partitions to consider are too unbalanced to admit any other tangle than the boring one that orients them all towards their large side,[21] the approach sketched in Section 9.1 offers a way forward. Let us illustrate this for the dual system, whose tangles of V were 'types of items'. Instead of trying to specify only those unbalanced partitions of S that have the form $v \in V$, consider the set V^* of all partitions of S. Define the *order* of $\{A, B\} \in V^*$ as

$$|A, B| := \sum_{a \in A} \sum_{b \in B} \sigma(a, b),$$

where

$$\sigma(a, b) = \left|\{\, v \in V : a(v) = b(v) = \vec{v}\,\}\right|$$

is the number of customers that bought both a and b. In the terminology of Section 9.1, this is the biased version of (O1). A partition of S gets high order, and is therefore disregarded when k-tangles are computed for small k, if it divides the contents $\vec{v}$ of many shopping baskets $v(S)$ – not the items missing from them – into two roughly equal parts.

This unifies the tangles of bespoke subsets of S that we considered in Section 6.3 into k-tangles for different k of the same set V^*.[22] For example, the set S' of green items in S will guide a tangle in $\vec{V^*}$ if there are enough ecologically minded customers in V. Indeed, partitions of low order will not divide S' into roughly equal parts: this would split the contents of the shopping baskets of many such customers evenly, generating lots of pairs (a, b) whose $\sigma(a, b)$ will count towards the order of such a partition. Hence most of S' will lie on the same side of – and thereby consistently specify – all partitions of low order.

Similarly, the set of the comparatively inexpensive items in S guides a tangle in $\vec{V}^*$. By contrast, the set of expensive items in S will not guide a tangle in $\vec{V}^*$: customers unable to afford them will not contribute to the value of σ for pairs of expensive items, while customers that can afford them are not necessarily likely to buy them more often than cheaper ones. Hence, $\sigma(a, b)$ will not be greater for pairs a, b of expensive items than for arbitrary pairs of items.

This pair of examples shows neatly that tangles in $\vec{V}^*$ deliver what we hoped for: they describe, loosely and without any need for difficult judgments in single cases as in standard clustering, groupings of items for whose purchase there exists some common motivation amongst the customers in V. In particular, they will do this for previously unknown kinds of motivation such as, perhaps, featuring a face on the packaging.

14.8 Text tangles: direct and indirect

In Chapter 6, Section 6.4, we discussed text tangles – in particular, 'tangles of words' in the literal sense that the partitions of a set V of texts whose tangles we might compute were given by individual words dividing V into the texts that contained this word versus those that did not – largely from a classification point of view. The idea was, for example, that a topic common to some of the texts in V could be identified by a set of words typical for that topic, and so tangles, ways of marking some of the words in V as 'typical', might identify such topics. Our discussion showed that this was unlikely to work when done naively. So let us see what additional tools we have available now to make it work.

We shall then look briefly at text tangles whose partitions of V in S can be arbitrary rather than having to correspond to single words. This will require us to restrict ourselves to k-tangles with respect to some order function. This is where we can build on existing work on text comparison and classification, but add a new and different twist.

Let us start where we left off in Section 6.4. We were looking at some hypothetical collection V of texts about hobbies, and tried to identify each of these hobbies by a tangle of words. These words, collected together in some set S, would be chosen by some pre-processing algorithm designed to filter out the most indicative words in a collection of texts.[23] If G, say, is the set of those words in S which we would intuitively associate with the hobby of gardening, our first hope was that the specification γ of S that specified every word in G as 'included' and all the others as 'not included' should form such a tangle identifying gardening. However we quickly saw that γ was unlikely to be a tangle.

At the other extreme, we found that G would likely have many small subsets G', perhaps consisting of only three words, that would define a tangle in this way: after all, every triple of words in a given text defines a tangle of agreement 1 focussed on that text. Our problem, then, was that we would get several 'gardening tangles' in this way, one for every such triple of gardening words. However there was another problem too: such tangles would exist not just for (many) sets S' of three words typical for some fixed hobby, but also for arbitrary triples S' of words from S. We would therefore get many meaningless tangles of agreement 1, but maybe none of higher agreement.

The underlying reason for these problems, but also their cure, are the same as in Section 14.7: the partitions of the set V of texts given by single words are too unbalanced. But we can consider instead the set S^* of all the partitions of V and study its k-tangles for increasing k, given some natural order function on S^* such as (O1) based on word counts. As always, we shall not be able to compute all these partitions, so we shall have to sample, but apart from this standard caveat we can then use the entire machinery of tangle algorithms outlined in Chapters 10 and 11.

More importantly, we can then also build on the fast-growing expertise in natural language processing, such as large language models. We can use this to assess the similarity between pairs of texts on which we built our order functions on S^* in Chapter 9. Many of these similarity measures depend not just on word counts but on other text parameters too. Those might be syntactic, or semantic in terms of known word fields, or based in unknown ways on neural networks trained to assess the semantic similarity of texts.

Rather than seeking to replace the work that continues to go into text analysis, our text tangles are designed to build on it. What they seek to replace is not the expert portion of this work, but the generic clustering that sometimes happens at the end of it and is just based on the similarities the AI has established.

14.9 Semantics and order functions

We discussed in Sections 2.3 and 5.3 how the notion of 'chair' can be captured as a tangle of potential features of furniture. For this chair example we considered as features only properties of furniture, such as being made of wood, which may or may not apply to a particular piece at hand. However when we try to train a computer to form or recognize notions, there is no reason to be so restrictive: we can also use other parameters that help us distinguish between different notions, such as context, or the time when it was fashionable.[24] And ultimately, as discussed in Section 6.5, we can try to capture notions by tangles of arbitrary partitions of a set of objects, or of basic predicates, that have no interpretation at all.

In Section 7.4 we saw how to use order functions to divide potential features into hierarchies, assigning low order to basic features and higher order to more specific ones. This is nowhere more relevant than when we employ tangles to capture notions in our ideas and perceptions: some notions are more fundamental than others and should have lower order.

In Chapter 9 we discussed generic order functions of set partitions, order functions that were defined in purely structural terms of our data. But when we try to teach a computer semantics, it is tempting to be more ambitious: we feel that some notions are better, or at least more useful, than others,[25] and we would like our computer to take this into account as it forms its own notions, or divides ours into more or less fundamental ones.

There are two quality criteria for notions that come to mind at once: one is relevance; the other what one might call clarity, or definiteness. Let us discuss these two a little, if only as a case in point.

As an example for relevance-based order we could choose to assign all colour questions high order in our search for furniture types, as long as we consider colour irrelevant to how furniture splits into types. As an example for clarity-based order, suppose we are trying to distinguish hairstyles. We are likely to get clearer answers if we ask when they were fashionable rather than whether they are pretty. So the former question should receive lower order on the clarity scale than the latter.

These two criteria, of relevance and clarity, may well conflict: being pretty or not is perhaps the most relevant aspect of a hairstyle, but it may also be the least clear in that people have divided opinions about it.

In Section 7.4 we discussed some pitfalls that came with defining order functions explicitly. But as we saw in Section 9.9, one can make even generic order functions aware of our preferences. Given some concrete criteria – in our example, relevance and clarity – this might require us to go through all the potential features s in our set S and assign them weights $w(s)$ to reflect their relevance or clarity – a truly daunting task.

However, we can also try to detect indicators for our criteria mechanically in our data and then use these, for example, in our choice of the similarity measure that defines an order function as in Section 9.1. In our example, both the relevance and the clarity of a question can be gleaned from how people answered that question compared with others. Questions that split the answer sets of many other questions roughly in half are, perhaps, less relevant on average than questions whose answer sets are more aligned with those of other questions.[26] And to measure the clarity of our questions we could simply add an option to answer 'don't know', and disregard questions where this option is chosen often.

What we have discussed so far is how a computer can learn the meaning of words as we use them, or come up with notions of its own that are formed by observing the world. Both these are aspects of what one would call the formation of passive vocabulary. Once this has been achieved, our computer will also want to know which of its tangle-encoded notions best describe a given object presented to it. This is a problem we discussed in Chapter 6.1: the problem of how to match a given object to a tangle of potential features, the tangle that best captures its actual features. See there for further discussion.

14.10 Acquiring meaning: an interactive thesaurus

Let us complement the previous section with a dream example of how tangles of notions might be applied, just to indicate what might be possible: let us build an interactive thesaurus for non-native speakers.

Roget's Thesaurus helps writers find the best word for what they are trying to say simply by grouping together some likely candidates. The relevant group of words can be found by looking up a word whose meaning is close to the intended meaning, but maybe does not quite capture it. If a word has multiple shades of meaning, it is linked to several such groups.

The value of this lies in offering the writer a relevant choice, but the thesaurus does not help with making a decision. This is fine if the writer knows all the words on offer, and perhaps just could not think of the right one. For learners of a language, however, this falls short of their need: they, too, will know what they are trying to express, but need help in finding the best word for it. Let us see how tangles can help them.

In the simplest model we could devise, for every word field currently offered together as a group of choices, a questionnaire S about aspects of the intended meaning of the word sought by someone consulting this word field: questions whose answers would allow an expert user to choose the right word from this field, or to determine that none exists.[27] These questions might ask about shades of meaning, the intended context, and so on – anything that helps us determine which of these words is the right one to choose. Since the situations in which a user intends to use a word are so varied, there will be many more possible sets of answers than words in this word field. So we shall not be able to match complete answer sets for S directly to the words in that field. But S could be designed in such a way that the words on offer match its tangles: we just have to build enough redundancy into those questions.

We will have to design all these S before our thesaurus is published, and test them on native speakers. Once they work, in that their tangles do correspond to the words on offer in that field, we can compute the tangle-distinguishing tree sets $T \subseteq S$ from Theorem 1. When the thesaurus is published and in use, it will be the questions from T, rather than the much longer list S, that are put to the user trying to identify the right word for their intended notion. Answering just these questions will steer them to the word that best fits their intended notion, because the tangles of S are determined by the locations of T.

Recall from Section 11.3 that the questions in T may be 'corners', combinations, of questions from S. So the questions a user really has to answer may be a little longer. But the subset of S needed to form the questions in T is still likely to be smaller than S itself. This is crucial for making a good thesaurus: it will matter to the user whether they need five questions or fifteen to be steered to the fitting word.

Devising a questionnaire S for each word field in the thesaurus, and answering all its questions for each word, may look like a lot of work. But this work has to be carried out only once, when the thesaurus is made: it is offset by a gain on the user side in that the questions asked are chosen specifically for each word field selected by the user, and are chosen particularly well for this word field.

In a more sophisticated model, we could grade the questions from S as more or less relevant for the corresponding word field, and assign them an order correspondingly. We would then obtain a hierarchy of tangles as described in Section 7.4, and obtain T from Theorem 2. The hierarchy of tangles of this world field would correspond to a hierarchy of more or less general or specific words in it, just as in reality.

For the task of grading the questions in S to assign them an order, our thesaurus might even learn from user interaction. Although a user may not know the meanings of the words on offer, they will be able to grade the questions put to them as more or less relevant to their specific search: tangles, after all, reflect notions, not words, and users come with such notions in their minds. Our tangles will then have to be recomputed from time to time when enough user feedback has been collected, and re-checked editorially against the words on offer. At this point, editors might also comment on user-generated tangles that do *not* have a corresponding word to match: such tangles will exist, since notions exist that are not exactly matched by words. Compared against the notions in the minds of speakers of other languages, however, this seems even more likely and worth addressing in ongoing editorial work.

Notes

Notes for the Preface

1. Part III assumes familiarity with some concepts covered in a first year university course in mathematics. Parts I and II can be read with no mathematical background.

2. The point here is that, since the minor is highly connected, it cannot be spread evenly across such bottlenecks: then the narrow bottleneck would separate two large parts of it, which would contradict its high connectivity. Thus, most of any highly connected minor will lie on the same side of any bottleneck. Compare Figure 7.1.

3. This is not unlike the situation we have in many real-world clustering problems. There, too, the essential information is often just *where* a big cluster lies; it may be of secondary importance, indeed distracting, to ask which points exactly should be considered as belonging to a given cluster. See Section 1.3.

Notes for Chapter 1

1. This section is deliberately written in abstract terms, to help us focus at this stage on the wide-ranging potential of tangles without the distraction of a particular example. We shall see lots of concrete examples in Chapter 4. Any reader who feels lost in too much abstraction here is invited to peek at the start of Chapter 4 now, or simply to read Sections 1.2 and 1.3 first. However, just to get a foot on the ground, think of 'phenomena' as earthquakes, or instances of sleeplessness.

2. Thus, the entire collection τ of hypothetical measurement readings is 'popular' with the elements of X.

3. In particular, these three readings are consistent in the usual sense that they *can* occur together. The reason we require this for all sets of up to three measurements, rather than just for any two, may be surprising but is immaterial at this level.

4. Indeed, consider the readings of any three measurements A, B, C specified by τ. If τ is typical for all the measurements we took in the popularity-based sense, with X as earlier, then on each of A, B and C less than a third of the phenomena in X disagree with τ (gave a different reading for this measurement than is specified by τ), so at least one phenomenon in X agrees with τ on all three of A, B and C. Hence τ is typical also in the consistency-based sense.

5. Make America Great Again; Donald Trump's 2016 presidential campaign slogan.

6. The partition of our set at the red bottleneck, for example, has the bottom cluster on one side and the other three clusters on the other side.

7. This is not to say that any definition of 'bottleneck partitions' and of 'consistency' will do: in cases where we feel we know roughly what the clusters should be, as in our example, the tangles resulting from our formal definitions should identify precisely these. In particular, there should be four tangles in Figure 1.2.

8. If desired, we can then use this to define consistent orientations of bottlenecks themselves after all: if we think of a bottleneck as the class of all the partitions at it, consistent orientations of bottleneck partitions will be well defined on these classes and thus induce orientations of the bottlenecks as such. But we shall not need this.

9. For example, we may later introduce another set V and assume with hindsight that $\vec{S}$ is the set of all sides of certain partitions of V. Then those partitions can be naturally denoted as s, with sides $\vec{s}$ and $\overleftarrow{s}$, so that in the end we have a set S of partitions of V for which our $\vec{S}$ is the set of all orientations of elements of S. Then everything will click into place at that point, and our notation will be much more natural than if we had denoted our set $\vec{S}$ generically as A when it first appeared.

Notes for Chapter 2

1. Logicians may prefer to say 'predicates' instead of 'features' here. That would be correct, but I am trying to avoid any (false) impression of formal precision at this informal early stage.

2. This may sound as though tangles were just new a clustering method after all, albeit one of features. The difference between tangles and clusters of features will become clearer in a moment. It is also addressed explicitly in Section 14.3.

3. Mathematically speaking, a *specification* of S is the image of any map $\sigma\colon S \to \vec{S}$ such that $\sigma(s) \in \{\vec{s}, \overleftarrow{s}\}$ for every $s \in S$, the subset $\sigma(S) = \{\sigma(s) \mid s \in S\}$ of $\vec{S}$.

4. It might seem more natural to say 'two' here, as in our wood/steel dichotomy above. Our definition of consistency is a little more stringent, because the mathematics behind tangles requires it; more on this in Section 7.3. Note that, formally, the elements of an inconsistent 'triple' need not be distinct: an inconsistent pair, two features $\vec{r}, \vec{s}$ not shared by any $v \in V$, also counts as an 'inconsistent triple', e.g. as the triple $\{\vec{r}, \vec{r}, \vec{s}\} = \{\vec{r}, \vec{s}\}$.

5. Here is a simple example. Suppose some of our furniture is made of wood, some of steel, some of wicker, and some of plastic. Denote these features as $\vec{p}, \vec{q}, \vec{r}, \vec{s}$, respectively, and assume that $S = \{p, q, r, s\}$. Then the specification $\tau = \{\overleftarrow{p}, \overleftarrow{q}, \overleftarrow{r}, \overleftarrow{s}\}$ of S is consistent, because for any three of its elements there are some items in V that have these three (negated) features: those that have the fourth of our original features. But no item in V has all four of the negated features in τ. We shall get back to this example in Section 2.4.

6. Let us ignore for the moment the possibility that the 'question' s may not have an ideal answer for chairs, as would be the case, say, for questions of colour rather than function. This is an issue we shall have to deal with, and which tangles can indeed deal with easily.

7. ... of which there are many: if S has 100 elements, it has 2^{100} specifications.

8. This is not to say that the use of tangles is free of all preconceptions, biases etc. For example, the choice of a survey S in the scenario from Section 1.2 is as loaded or neutral as is would be in any other study that starts with a survey. The statement above is meant relative to the given S once chosen. In Chapter 7 we shall discuss how the deliberate use of preconceptions, e.g. by declaring some questions in S as more fundamental than others, can help to improve tangles based on such preconceptions. We shall also see how to do the opposite: how to find tangles that arise naturally from the raw data of S and V, without any further interference from ourselves.

9. This notation anticipates our mathematical definition, to be given in Chapter 7, of 'specifications' as (the images of) functions $\sigma\colon S \to \vec{S}$ that assign to every $s \in S$ either $\vec{s}$ or $\overleftarrow{s}$.

10. In our formal definition of tangles in Chapter 7 we shall always identify a feature with the set of elements of V that have it. Then $\vec{r}, \vec{s}, \vec{t}$ will denote subsets of V.

11. Consider as S the set of partitions whose dividing line is no longer than the handle is thick. Then all the partitions whose dividing line cuts vertically through the handle lie in S, and hence τ has to orient them. Clearly, τ cannot orient two such partitions 'away from each

other', so that their sides chosen by τ are disjoint, since this would make τ inconsistent. But neither can they point towards each other. If they did, we could cut up the space between them, which is part of the handle, into singleton elements of V by similar such partitions in S, just as we cut up the entire set V into singletons in our earlier proof. If singletons $\{v\} \subseteq \tau$ are ruled out by our choice of $\mathcal{F}$, which we assume, we obtain the same contradiction as earlier.

12. Unlike the principal and the focussed tangles, 'black hole' tangles are not a rigorously defined type of tangle, because the notion depends on what we regard as 'inuitive clusters'. Let us remember this whenever we refer to them later in the book.

13. More precisely, if $p > \frac{1}{2}$ is maximum with $\frac{\sum\{w(v) : v(s)=\tau(s)\}}{\sum\{w(v) : v \in V\}} \geqslant p$ for every $s \in S$, then w *guides* τ with *reliability p*.

Notes for Chapter 3

1. This will be our first example, in Section 4.1, of an application of tangles. Here, V is a set of patients with known records, S consists of the full range of possible diagnostic measurements and their combinations, and features $\vec{s}$ correspond to the possible readings of the measurements $s \in S$.

2. As indicated earlier, we may have enriched the set $\vec{S}$ by adding some combinations of features from our original $\vec{S}$ to make it rich enough to apply our tangle theorems. The elements of the enriched S are now combinations of questions. But for simplicity we refer to these, too, as 'questions'.

3. We are talking about 'predictions' here at a precision level of weather forecasts; not about absolute predictions whose failure, even once, invalidates the entire theory.

4. Thus, 'holding the view $\vec{s}$' would count as an 'action' in our new context; if desired, think of it as the action of answering the question s as $\vec{s}$.

Notes for Chapter 4

1. Measurements with more than two, but finitely many, outcomes can be modelled as several measurements of this simple 0/1 kind. We shall discuss this later.

2. It may work better in controlled environments, such as diagnosing the cause of a computer crash rather than of an illness of a human being.

3. So testing for an abnormally slow pulse would be considered as another measurement. Similarly, there could be one measurement testing for body temperature in the range typical for viral infections, and another for bacterial infections, although these can be taken together in the same physical measurement of body temperature.

4. More precisely, we require of S that if a positive reading $\vec{s}$ is associated with an illness then more than two thirds – but by no means all – of the patients with this illness have this positive reading $\vec{s}$, and that for all other measurements in S we have fewer than a third false positives. Then the set X of patients having the illness will guide τ in the sense of Section 2.5, making it a tangle at least for $\mathcal{F} = \mathcal{F}_1$.

5. In the 5′ to 3′ direction, say. We also assume that our molecules are aligned, which is possible only if they come from sufficiently similar or related organisms. In Section 12.3 we discuss how to use tangles for unaligned DNA and protein clustering.

6. In other words: either they are equal or each is equal to the inverse of the other.

7. A multiset is a set that can contain several copies of the same element. If some organism left five DNA molecules in our sample, say, and these gave rise to five identical sequences, we want V to contain this sequence 'five times'.

8. Remember that the inverse of a sequence is taken not just by reversing it but also by replacing every base with its partner base.

9. Hence, just like the two strands of a given DNA molecule, these two tangles describe the same thing: we might just pick one of them, knowing that the other can be recovered from our pick by inversion. However, since we have no way of specifying a default strand in a DNA molecule without first identifying the species which it represents (and then checking it against a database), which is part of what we are hoping to use tangles for, we had to keep both strands. This now shows in that our tangles will likely come in pairs, but we may discard one of each pair if desired.

10. Think of molecular composition, or perhaps shape; they do not have to be of the same type.

11. See Sections 6.1 and 14.1 for more on how to find among the tangles of S one that best represents the features $v(S)$ of a given element $v \in V$.

Notes for Chapter 5

1. Questions with more than two possible answers can be modelled as several questions of this simple yes/no type. We shall discuss this in more detail in a moment.

2. Still, knowing the tangles of S can also help us determine such groups. We shall discuss this in Sections 6.1, 14.1 and 14.2 in the context of tangle-based clustering.

3. For example, if a question s in S asks for a numerical value between -5 and 5 on a scale from 'do not agree at all' to 'agree entirely', we can replace s with eleven imaginary yes/no questions s' asking whether the value of s' is $-5,\ldots,5$, respectively, or with the ten yes/no questions s' asking whether the value of s' is less than i, for $i = -4,\ldots,5$.

4. This is not the same as saying that all possible specifications of S will be tangles. Specifications that contain an inconsistent triple will be inconsistent even for $\mathcal{F} = \emptyset$, hence not a tangle. But these specifications are not among those returned in the poll.

5. For example, we might be interested in the opinions about hooligans among football crowds, but want to exclude hooligans themselves from this survey. Since we may not be able to identify them when we hand out the questionnaires, but know some answer patterns they are likely to give, we can add these patterns to $\mathcal{F}$ to ensure that the tangles found are mindsets of spectators that are not themselves hooligans.

6. Alternatively, the dictionary might give us a list of predicates such that a given thing is a chair if and only if it satisfies at least one of these, or a mixture of the two.

7. In fact, Wittgenstein's families are best represented by tangles of the dual feature system introduced in Section 7.7: for every few members of a family, such as every three, there are many features – e.g., at least some $n \geqslant 1$ – which those three members share, but not all the family members have to have all those features.

8. This was defined in Section 2.3.

Notes for Chapter 6

1. This notation does not determine which of the sides A and B is $\vec{s}$ and which is $\overleftarrow{s}$.

2. How these can be captured in precise mathematical terms, without reference to clusters, will be a matter for Chapter 9.

3. If desired, we could give an ad hoc definition of 'bottleneck partitions', just for this example, that makes no reference to point clusters. For example, let $\mathcal{P}$ be the set of polygonal lines in the figure, between boundary points, whose vertices lie in V. We might then take S to be the partitions $\{A, B\}$ of V for which there exists a line $\ell \in \mathcal{P}$ of length at most some constant k, something like the width of the handle, such that we cannot connect a point in A to a point in B by a line in $\mathcal{P}$ that does not cross ℓ. With such a definition, clumsy as it is, we could *prove* that each of the four point clusters lies mostly on one side of

every partition in S, and thus defines an orientation of S – indeed a consistent one, a tangle.

4. Recall that, in Section 2.3, we thought of the tangle capturing the notion of a chair as an 'ideal chair', an additional 'virtual element' of V.

5. If $\tau_v = \{\, v(s) : s \in S \,\}$ is a tangle, then $\tau_v^+ = \tau_v$.

6. We consider tangles with agreement $n = 2$ here.

7. The points in the pouch do form a natural cluster, which can even be detected as X_τ. But only if we raise our threshold to its theoretical maximum of $|S|$.

8. The sets $\overline{s}$ with $s \notin \{s_1, s_2, s_3\}$, which guide the other three tangles τ_4, τ_5, τ_6 and form their corresponding sets $X_{\tau_4}, X_{\tau_5}, X_{\tau_6}$, clearly are natural clusters. Note that we can make them as big as we like without invalidating the example.

9. As standards for S decrease, regions get subdivided into smaller regions, because more partitions are allowed. This appears to contradict the way in which the regions in Figure 6.4 are nested, which appears to be the wrong way round: the red line is the clearest but appears to bound the smallest region, followed by the green line which bounds a larger region, while the blue line, which is the least clear, bounds an even larger region. But this seeming inversion of nestedness is an artefact of comparing only three partitions: note that the *outside* area of the red circle is partitioned further by the green circle, whose outside is in turn partitioned by the blue line.

10. This is not the same as just a group of items 'often bought together', as are already suggested by online shops today. Unlike those, a hypothetical shopping basket defined by a tangle is also typical, for example, in what it does *not* include. Note, however, that tangles are typical for shopping baskets only in terms of what they do or do not include; they need not be realistic in other ways. For example, they might include half of all the items in the shop. See below for examples.

11. This is analogous to thinking of the chair tangle as an ideal chair, or of a mindset tangle as a hypothetical person holding such views. In each case, the tangle is a set of features typical for the elements of V, and we may think of it as a hypothetical 'typical' element of V that has exactly these features.

12. For mathematicians: there is a formal duality here in that $v(s) = s(v)$ with the obvious interpretations of v as a function $S \to \{0, 1\}$ and s as a function $V \to \{0, 1\}$. The customer preferences described in two ways can be formalized as the edge set of the bipartite graph with vertex classes V and S and edges vs whenever v bought s. This edge set can be described alternatively as a list of neighbourhoods of the vertices in V or of those in S, two dual ways of describing the same graph.

13. In our current informal setup, these would be tangles of subsets of S, not of S itself – for example, of the subset of the ecologically critical

or the unusually priced items. When we revisit this example in Section 14.7, we shall find a way of capturing all groups of items, such as the ecological or the inexpensive ones, by tangles of the same set.

14. Unlike in other contexts, every v in this example has a default specification $\vec{v}$: the set of purchases of v rather than his non-purchases. The set $\{\,\vec{v} \mid v \in V\,\}$, therefore, is well defined.

15. Indeed, more is true: for every three customers v, regardless of how the tangle τ specified them, there exists a set of at least n items that are each bought by v if $\tau(v) = \vec{v}$ and not bought by v if $\tau(v) = \overleftarrow{v}$.

16. This phrase is based on a fortuitous quirk of our tangle terminology. When the elements s of S correspond to words, each dividing our set V of texts into the set $\vec{s}$ of those texts that contain the word s and the set $\overleftarrow{s}$ that do not, then the tangles of S are quite literally 'tangles of words'.

17. This was different in our chair example in Chapter 2, where any sizeable collection of chairs would have at least one chair, indeed many, that has four legs, a flat surface to sit on, and a nearly vertical back – and similarly for any other triple of features by which chairs are commonly identified.

18. Unlike, for example, in mathematics: there, conscious decisions about which properties of the (mathematical) objects studied should be bundled into notions to facilitate further study are part of our daily bread. And they are as important as elsewhere in life, since they determine what structures of the objects under study become visible and can therefore be explored.

19. For example, the computer might learn by showing us pictures of groups of objects and asking which of them 'do not belong' in this group.

20. Despite the wording here, this is not entirely a matter of taste. Very crudely, one might say that good notions facilitate the expression of theories that describe the world better than others – for example, in the sense of making more specific or substantial predictions.

Notes for Chapter 7

1. Ready-to-use software to support this is available at [1].

2. In the terminology of this book we would call this the definition of *consistency* for orientations of all the separations of order less than k in the given graph.

3. It is meaningless here to ask which is which: whether $\vec{s} = A$ and $\overleftarrow{s} = B$, or $\vec{s} = B$ and $\overleftarrow{s} = A$. This is because the two elements of $s = \{A, B\}$ are 'born equal': the expressions '$\{A, B\}$' and '$\{B, A\}$' denote the same set, or unordered pair. If this did not worry you, let it not worry you now.

4. Thus, a *features system* is just a set of subsets of a given set V that is closed under complementation: for every $A = \vec{s} \in \vec{S}$ also its complement $\overleftarrow{s} = \overline{A}$ lies in $\vec{S}$.

5. These features $\vec{r}$, $\vec{s}$, $\vec{t}$ need not be distinct. For example, our subset of $\vec{S}$ will already be inconsistent if it contains two features $\vec{r}$, $\vec{s}$ such that $\vec{r} \wedge \vec{s} = \emptyset$; this is included in the definition given, simply by taking $\vec{t} = \vec{s}$ or $\vec{t} = \vec{r}$.

6. Technically, one can replace our requirement of consistency in the definition of a tangle (see below) with the weaker requirement that tangles do not contain inconsistent triples just of this particular form: all the tangle theorems of Chapter 8 remain true for these more general tangles. In the literature they are known as *profiles* [13].

7. In classical clustering, these pre-tangles are sometimes called 'clusters in the feature space': sets of features $\vec{s} \in \vec{S}$ that are pairwise 'close' in that any two of them are common features of many $v \in V$, those in the pie segments in which they overlap. Let us note at this point that tangles are something more subtle than such 'clusters in the feature space': those are pre-tangles in our terminology, not tangles. See Section 14.3 for more.

8. For example, if a tangle of S_i identifies the chairs in our set V of furniture, we might find that S_k for $k = i + 1$ has three tangles that induce this chair tangle of S_i: one for armchairs, another for dining chairs, and a third for garden chairs.

9. Formally, the term 'order function' does not denote any particular mathematical type of function, but is applied to any real-valued function on S that is intended to be used to define a hierarchy $S_1 \subseteq S_2 \subseteq \ldots$ of subsets of S as discussed here. If we start with such a hierarchy of sets, we can define its corresponding order function retrospectively, by assigning order 0 to all the elements of S_1, order 1 to all the elements of $S_2 \smallsetminus S_1$, and so on as $k = 1, 2, \ldots$.

10. From now on, we shall assume in all examples where S 'is a questionnaire' that the people in V have already answered all its questions. Only then do the elements of S become potential features in our new formal sense, i.e., partitions of V: these partitions are, for every 'question' s, given as the sets A and B of the people in V that answered s as yes or as no, respectively, so that s becomes the partition $\{A, B\}$.

11. There are actually more partitions in S_k, which split off just a few points; those are all oriented towards the side that contains most of the points. See Note 11 in Section 2.4.

12. Figure 14.4 offers a more elaborate example in a clustering scenario that shows the evolution and a hierarchy of tangles similar to Figure 7.4.

13. For example, $\overleftarrow{r} \wedge (\vec{r} \vee \vec{s})$ and its inverse are Boolean expressions of $\vec{r}$ and $\vec{s}$.

14. And those that are not can often be made 'locally submodular' by adding some corners on the fly, as our algorithms need them.

15. Then the inverse $-f$ of f is associated with the subset $(-f)^{-1}(1) = f^{-1}(-1)$ of V, the complement of the set $f^{-1}(1)$ with which f is associated.

16. Thus, although our features themselves are no longer binary in that they map V to more than two possible values, they still come in pairs.

17. As previously, it makes no sense to ask which is which: we could have $\vec{s} = f$ and $\overleftarrow{s} = -f$, or $\vec{s} = -f$ and $\overleftarrow{s} = f$.

18. In the case of binary feature systems, i.e. of set partitions, this was not necessary, since for any pair $\vec{r}$, $\vec{s}$ of orientations of nested partitions r and s one of $\vec{r} \vee \vec{s}$ and $\vec{r} \wedge \vec{s}$ would be equal to $\vec{r}$ or $\vec{s}$ or their complement, which were all in $\vec{S}$ by assumption, or equal to $\emptyset$ or V. As $\{\emptyset, V\}$ was not considered to be a partition of V, these latter could not formally be in $\vec{S}$, but this did not matter since the 'degenerate partition' $\{\emptyset, V\}$ does not cause any problems when submodularity is applied. In our new setting, however, $\vec{r} \wedge \vec{s}$ and $\vec{r} \vee \vec{s}$ can be interesting features whose existence in $\vec{S}$ matters but does not follow from $r, s \in S$, even when they are nested, and is therefore required by submodularity. For example, two positive functions f and g will *always* be nested, because $-f \leqslant g$, but neither their infimum nor their supremum need be equal to any of $f, -f, g$ or $-g$.

19. This special case of consistency, the *profile* property from [13], is the only consistency aspect about tangles that we need in the premise of our main tangle theorems.

20. Other ways are indicated in Section 10.2.

21. For mathematicians: E is the edge set of a bipartite graph with vertex classes X and Y.

22. For mathematicians: this is well defined, stated symmetrically as above, since everything we have said has been symmetrical in X and Y.

23. The 'special relationship' exists between elements $v \in V$ and $s \in S$ for which $v \in \vec{s}$ and, equivalently, $s \in \vec{v}$.

Notes for Chapter 8

1. Remember that tangles can capture clusters also when these are 'fuzzy'. They locate even such fuzzy clusters in V precisely, by orienting all the partitions in T towards where most of such a cluster lies. This does not require, however, that all the *points* of that cluster also have to lie in this precisely specified location of T. In this way, tangles can make the location in T even of a fuzzy cluster precise without attempting the impossible, to specify where each individual cluster point lies.

2. To make the picture more intuitive, one can deform the disc so as to make the dividing lines shorter, as in Figure 8.1. When one does that, the shape of V will begin to resemble a tree. But this is only a matter of display: the crucial mathematical property of a tree set is that it can displayed like this at all, with V either round or tree-shaped.

3. It follows, for example, from the representability of finite tree sets by the edge sets of graph-theoretical trees and the fact that every $v \in V$ induces a tangle of T.

4. Pre-tangles were defined in Section 7.3: they are orientations of a set of partitions that contain no inconsistent pair but may contain inconsistent triples. All tangles are also pre-tangles, but sets of partitions, even tree sets, can have pre-tangles that are not tangles.

5. In the case of advanced feature systems (Section 7.6), minimality refers to the partial ordering $\leqslant$ that comes with the feature system.

6. If $\vec{t} \in \tau$, say, then $\vec{t} \supseteq \vec{s}$ for some $\vec{s} \in \sigma$. Then $\overleftarrow{t}$ cannot lie in τ, because $\vec{s}$ does but $\overleftarrow{t} \cap \vec{s} = \emptyset$. Similarly for advanced feature systems.

7. Indeed, if $\tau \neq \tau'$ then T contains a partition $\{C, D\}$ of V such that $C \in \tau$ and $D \in \tau'$. Then $V_\tau \subseteq C$ is disjoint from $V_{\tau'} \subseteq D$, since $C \cap D = \emptyset$.

8. As Figure 8.1 suggests, the tree shape of a tree set T of partitions of V can be made visible by a graph-theoretical tree whose nodes correspond to the locations of T and whose edges correspond to the elements of T. These elements may then be viewed as partitions not only of V but also of the set of nodes of this graph-theoretical tree. The locations of T, which by definition are subsets of T, then correspond to the edge sets of the graph-theoretical stars at the nodes of this tree [8].

9. Sets σ of advanced features (Section 7.6) are *stars* if $\vec{r} \geqslant \overleftarrow{s}$ for all $\vec{r}, \vec{s} \in \sigma$.

10. Indeed, distinct elements of a location σ of a pre-tangle τ of a tree set cannot contain each other, because they are both minimal in τ. For the same reason, their inverses cannot contain each other. They cannot be disjoint, because they would then form an inconsistent pair in τ. Since they are nested, the only possibility that remains is that their inverses are disjoint: that they form a star.

11. This means that they specify some element of T differently, and therefore do not induce the same tangle of T. But distinct tangles of T live at different locations.

12. There can even be several empty locations, but no more than non-empty ones. In graph-theoretical terms: locations at leaves of our tree (set) are never empty, and trees have more leaves than branching nodes.

13. Indeed, let σ be any location of T. Pick an element $\vec{s}$ of σ. If T is minimal with its property in Theorem 1, then S has distinct tangles τ, τ' that agree on $T - s$ but differ on s. Let τ be the one that contains $\vec{s}$.

Then τ' contains $\overleftrightarrow{s}$, and hence also $\sigma \setminus \{\vec{s}\}$: every $\vec{s'} \in \sigma \setminus \{\vec{s}\}$ satisfies $\overleftrightarrow{s'} \cap \overleftrightarrow{s} = \emptyset$, because σ is a star, so $\overleftrightarrow{s'}$ cannot lie in τ', because $\overleftrightarrow{s}$ does. But τ' agrees with τ on $T - s$, and hence also on $\sigma \setminus \{\vec{s}\}$. Thus, $\sigma \subseteq \tau$. Since $\tau \cap \vec{T}$ is a tangle of T, it is the unique pre-tangle of T living at σ. Hence τ is a tangle of S that lives at σ.

14. To strain our black hole analogy a bit further: if we choose T as S, our three celestial clusters can still be arranged around a void centre, but doing so is merely a way of organizing them that we impose to display their symmetry. The empty central location has no 'mass' now: not only is it empty as a subset of V, but no tangle of S lives there.

15. More on this, including a precise definition of 'canonical', can be found in [4, 5].

16. In fact, it has exactly $\ell + 1$ pre-tangles. This is because the pre-tangles of a tree set T correspond to the nodes of a graph-theoretical tree with $|T|$ edges [8], and a tree with ℓ edges has exactly $\ell + 1$ nodes.

17. See Chapter 11, Section 11.3.

18. Such an s will always exist for distinct tangles of the same order, i.e., for distinct tangles of the same set S_k. But a tangle of order k is not distinguishable from the tangles of order $\ell < k$ that it induces on the sets S_ℓ. Our assumption that s distinguishes two tangles, of orders k and ℓ, say, thus means in particular that $|s| < \min\{k, \ell\}$, and it implies that the tangle of higher order does not induce the tangle of lower order.

19. In this book, unless otherwise mentioned or apparent from the context, *partitions* are into exactly two non-empty sets.

20. Maximal in terms of inclusion as subsets of $\vec{S}$.

21. Formally: they live at the locations of $T \cap S_k$ that correspond to the maximal graph-theoretical stars of the tree $\mathcal{T}_k$, the sets of edges at a given node of that tree.

22. It is easy to check that a given tree set that is claimed to be over $\mathcal{F}$ is indeed over $\mathcal{F}$: we just have to compute its locations, which is easy, and check that they are all in $\mathcal{F}$.

Notes for Chapter 9

1. For example, if an order function on the partitions of V sums certain numbers associated with the pairs u, v of elements of V, e.g. as our simplest order function (O1) does, we would normalize it by dividing it by the total number $|V|^2$ of such pairs. But we might also count similarities on a logarithmic or exponential scale, say, depending on the source and type of our data.

2. In Section 1.3 we called such s the *bottlenecks* of V. Similarity was then given naturally as proximity in the picture referred to.

3. From a clustering perspective it is important to note that we only need to define *pairwise* similarity in terms of $\vec{S}$ here: not 'groups of similar objects', which would be clusters in the traditional sense and may be difficult to pinpoint.

4. Here is a sketch of the easy proof. Given $\vec{r} = A$ and $\vec{s} = C$, the sum $\sigma(A)$ of all $\sigma(u, v)$ with $u \in A$ and $v \in \overline{A}$ is $|r|$, and the analogously defined sum $\sigma(C)$ is $|s|$. Now computing these sums for $\vec{r} \vee \vec{s} = A \cup C$ and $\vec{r} \wedge \vec{s} = A \cap C$, we notice that the terms $\sigma(u, v)$ that count towards $\sigma(A \cup C) + \sigma(A \cap C)$ also count towards $\sigma(A) + \sigma(C)$, and equally often. Thus, $|\vec{r} \vee \vec{s}| + |\vec{r} \wedge \vec{s}| = \sigma(A \cup C) + \sigma(A \cap C) \leqslant \sigma(A) + \sigma(C) = |r| + |s|$ as desired.

5. It is important to use -1 rather than 0 here. Indeed, if f encodes the orientation A of a partition $\{A, \overline{A}\}$ of V by sending A to 1, we want $-f$, the inverse of f as a feature, to encode the inverse partition $\overline{A} = V \smallsetminus A$ by sending $\overline{A}$ to 1. It does that if f sends $\overline{A}$ to -1, but not if f sends $\overline{A}$ to 0.

6. As earlier, the assumption is needed to make the order function submodular.

7. Recall that in our earlier binary setup the yes/no questions in S could be interpreted as partitions of V, but we used them to define order functions not just on S but on the entire set of partitions of V. In our advanced setup now, the questions are once more the same kind of things as the features we are considering: both are $V \to \mathbb{R}$ functions. And, as before, we use the functions in Q to define – in various ways – the order $|f|$ of an arbitrary function $f : V \to \mathbb{R}$, one of our features. In particular, these f need not have the same range in $\mathbb{R}$ as the functions in Q that are our questions.

8. It is perfectly acceptable to model the yes/no answers received from the people in V to the questions in Q by having $q \in Q$ send $v \in V$ to 1 if v answered q as yes, and to 0 if it answered no. Then q does not encode an oriented partition of V whose inverse, as a partition, corresponds to the inverse of q as a feature, the function $-q$. But there is no reason it should. If we want Q to be an advanced feature system that is the same as – in mathematical terms, isomorphic to – the feature system of the partitions of V it defines, we encode its answers by 1 and -1 rather than 1 and 0.

9. This is analogous – or in mathematical terms, dual – to our earlier observation that $\langle f, g \rangle$ measures the correlation between f and g viewed as $V \to \mathbb{R}$ functions.

10. In some contexts it is desirable that the similarity between $u, v \in V$ should depend only on how the difference of their views is distributed over the questions in Q, but be independent of their strengths. In this case one can define the similarity between u and v as $\sigma(u, v) :=$ $\langle u, v \rangle / (\|u\| \|v\|)$, where $\|u\|$ and $\|v\|$ denote the Euclidean norm, or 'length', of u and v viewed as vectors in $\mathbb{R}^Q$. This $\sigma(u, v)$ does not

change when we replace u and v with αu and βv for positive real numbers α and β. It can be interpreted as the cosine of the angle between these vectors and is known as their *cosine similarity*.

11. This is an aspect specifically of vector spaces that makes them simpler than other algebraic structures. The use of algebraic structures to determine the order of a partition is in no way limited to vector spaces; these just provide a particularly ubiquitous and simple example.

12. Graph theorists will recognize in N the adjacency matrix of the bipartite similarity graph with vertex classes A and B.

13. The rules for adding and multiplying the elements 0 and 1 with each other are the same here as for integers, except that $1 + 1$ is defined to be 0. For $n = 3$, for example, we have $(0, 1, 1) + (1, 1, 0) = (1, 0, 1)$.

14. Although our definition of $|s|$ depended on which side of s is A and which is B, the result in fact does not: swapping A and B yields the same value of $|s|$. This is a fact from linear algebra that is not obvious; in our case it says that $\operatorname{rank} N = \operatorname{rank} N^{\top}$.

15. In mathematical terms: it can happen that the columns indexed by $P \smallsetminus \{p\}$ generate the columns of M_A when restricted to A and the columns of M_B when restricted to B, but by different linear combinations.

16. Matroid theorists will recognize this as the connectivity function of the linear matroid of the rows of M.

17. This is more likely the case in science applications, where Q is a set of measurements rather than questions, but for readability we shall stick to our interpretation where Q is a set of 'questions' answered by the 'people' in V.

18. Recall that every $v \in V$ is such a function; its values $v(q) = q(v)$ are displayed in the row of M indexed by v.

19. For scalability.

20. In particular, $\det J_A \geqslant 0$, since these volumes are non-negative.

21. This is not a severe restriction in our context if $|V| \leqslant |Q|$: recall that we began to study this order function precisely because, with real-valued data, it happens so often that all the vectors $v(Q)$ are linearly independent – even if they have large subsets of small volume.

22. Recall that we always measure the $|A|$-dimensional volume of this parallelotope.

23. Adding a to A increases $|A, B|$ by $2 \log h(a, A) - 2 \log h(a, V)$; see [25]. In the rare case that $h(a, A) = h(a, V)$, this leaves $|A, B|$ unchanged.

24. Recall that $\det J_A$ is the product of all the eigenvalues of J_A.

25. Graph theorists will recognize W_+ as a weighted incidence matrix of the complete graph on V.

26. In the case of L, both determinants are non-zero as soon as the graph on V with edge set $\{ uv \mid \sigma(u, v) \neq 0 \}$ is connected.

27. Everything in this section comes from discrete probability and can be expressed more generally and precisely in probabilistic terms. However, probabilistic terminology, though highly intuitive, can also be very misleading when used informally without a rigorous and known mathematical background. Since we assume no such background in this book, we avoid its use except for informal analogy.

28. Formally, this is our surprise at seeing the answer (x', x'') given for the map $f: V \to X \times X$ that sends each $v \in V$ to the pair $(q'(v), q''(v))$ of its answers to the questions q' and q''.

29. Indeed, by our definition of independence, $p_f(x', x'') = p_{q'}(x') \cdot p_{q''}(x'')$. So the surprise at seeing f answered as (x', x'') will be $1/p_f(x', x'') = 1/(p_{q'}(x') \cdot p_{q''}(x''))$, which is the product of the surprises $1/p_{q'}(x')$ and $1/p_{q''}(x'')$ at seeing q' answered as x' and at q'' answered as x''.

30. For example, if the same surprise happens to us twice, we are 'twice as surprised'.

31. If we change $|V|$ by some factor but keep the relative sizes of the sets $f^{-1}(x)$ the same, then the sizes of these sets change by the same factor. Hence the numbers $p_f(x)$ remain unchanged, and so does $H(f)$.

32. Formally, by definition of independence, the relative sizes of the sets $f^{-1}(y)$ between $y = $ yes and $y = $ no for $f = q''$ do not change when we restrict f to the set of $v \in V$ that answered q' as yes, or to those that answered q' as no. Hence the second sum in the definition of $H(q''|q')$ equals $H(q'')$ for every x, and averaging these second sums over all x as done in the first sum has no further effect.

33. We should not forget that, intuitive though it may be, this 'help' or 'telling us' about the answers to f_2 once we know f_1, has a formal definition here that takes no recourse to any interpretations of f_1 and f_2, e.g. to what they mean if they are questions from Q. The same goes for the notion of 'mutual information' we are about to introduce.

34. We used to think of v itself as the map on Q recording all its answers. However, it is important now to count how many $v \in V$ gave a particular set of answers, so we need to distinguish it notationally from this set: the map Q on V need not be injective.

35. This is perhaps more obvious for technical applications, where Q consists of measurements made of 'objects' in V than in a narrow questionnaire setting. If questions are offered with an option to choose 'don't know' or 'don't care' as an answer, then these answers, of course, will be chosen more often by people not interested in the subject of the question. In our formal setting many people choosing this answer counts as 'structure' that reduces the entropy. Only in the interpretation will it then become visible, that this 'structure' may not be interesting – such as a tangle choosing 'don't care' as all the answers to Q.

36. This is not to say that they cannot carry useful information. Indeed, although they do not directly correspond to any obvious clusters in our

dataset, they can carry structural information about those tangles that do – for example, that the clusters are not organized into clusters of clusters. See Section 14.1.

37. In particular, there is no tangle that includes $\{\bar{p}, \bar{q}, \bar{r}, \bar{s}\}$: this tangle would also have to orient the partition $t = \{A \cup B, C \cup D\}$, but both orientations of t form an inconsistent triple with two features from $\{\bar{p}, \bar{q}, \bar{r}, \bar{s}\}$.

Notes for Chapter 10

1. This is computationally fast if the edge weights are non-negative, and only then.

2. This is a mathematical term coined over a hundred years ago, which in our context simply means 'related to the eigenvalues of a Laplacian matrix'. A number λ is an *eigenvalue* of a matrix A if there exists a vector $f \neq 0$ such that $Af = \lambda f$. There may be more than one such f; they are the *eigenvectors* for the eigenvalue λ. Eigenvalues and their eigenvectors play a key role in linear algebra. But no understanding of their theory is required here: the few properties we need will be stated explicitly.

3. Spectral clustering will feature again in Section 14.5, where we compare it with tangle-induced clustering. When the latter is based on a spectral feature system such as discussed below, this comparison is particularly instructive.

4. In graph-theoretical language: we assume that the 'similarity graph' on V whose edges are those $\{u, v\}$ on which σ is nonzero is connected. If it is not, we work with its components separately, find their tangles, and combine them all at the end.

5. For the last equality in $(\overline{O}7)$ use that $u^{\top} I u = -(Mu)^{\top} Mu$ by definition of I.

6. The principal components of a vector space of real-valued functions can be thought of as representing as much of their variety as is possible in few dimensions. It is a standard technique in data analysis to represent the actual data, in our case the functions $q: V \to \mathbb{R}$ or $v: Q \to \mathbb{R}$, by their projections to a subspace spanned by their first few, here ℓ, principal components, in order to make the size of the dataset more manageable.

Notes for Chapter 11

1. Even well-designed questionnaires can include questions that are 'irrelevant' in our sense. This is because the relevance of a question in our sense is determined not in terms of its meaning but quantitatively by how the entire questionnaire was answered. Indeed, whether or not a question is relevant in a given context may well be a main result of the study we are analysing; recall the handedness example discussed in Section 9.9.

2. Section 14.4 in Chapter 14 addresses the difference between hierachical and tangle clustering.

3. The algorithm needs to add corners only to make our tree of tangles nested, not to distinguish more tangles: if all we require is a *small* set $T \subseteq S$ that distinguishes all the tangles of S or in $\vec{S}$, and we do not mind if elements of T cross, we can easily compute T in other ways without ever adding corners.

4. Recall that we are never interested in tangles of S^* itself, because the only such tangles are the focussed tangles τ_v. But we are talking about k-tangles of S^* here, and these are more relevant than tangles of proper subsets of S_k^*.

5. For example, if we have just checked that a certain triple $\{A, B, C\}$ is consistent, or not in $\mathcal{F}_n$, we can infer from $A' \supseteq A$ that $\{A', B, C\}$ is consistent or not in $\mathcal{F}_n$ too.

6. To see this, rewrite $\wedge$ as $\cap$: clearly, $A \cap B \cap (A \cap B) = A \cap B$.

7. For instance, recall the pie example from Section 7.3. The set S there consisted of n pairwise crossing partitions which divided the set V, drawn as points in a disc, into $2n$ segments. Taking corners of elements of S, and corners of corners etc, will generate at most 2^{2n-1} partitions, which is likely to be much smaller than the number $2^{|V|-1}$ of all partitions of V.

8. The figure shows an example with $\ell < k$ in which neither ρ nor τ orients s. If they do, e.g. when $\ell = k$, we may still assume that r and s do not distinguish each other's pair of tangles: otherwise we could delete r or s from T. We can then find opposite corners of r and s that are both home to one of the four tangles. If our order function is submodular, then either one of these opposite corners has order at most ℓ and can replace r as a ρ, τ-distinguisher, or it has order at most k and can replace s as a φ, ψ-distinguisher in T.

9. More precisely, we use a secondary submodular order function just to break ties when our chosen order function assigns two partitions the same order. The combination of the two order functions will be submodular as soon as the first is.

10. It is of theoretical significance for computational isomorphism problems. This takes us too far afield; readers interested are encouraged to consult [24].

11. In particular, it does not stop by 'running out of steam' because every feature it wants to add to L is already in L, but it hasn't added a pair of inverse features to L.

12. The pie example can be used to show this; see the footnote after next.

13. This will happen no matter how we placed those arrows on the edges of T. This is because graph-theoretic trees, by definition, have no cycles (around which those arrows might send us forever as we follow them).

14. Consider the pie example of Figure 7.3. Let S consist of the three partitions of V marked by the diagonals, and let n be just large enough that each of the six segments contains fewer than n points. Then any $\mathcal{F}$-tangle τ of S must orient all the diagonals clockwise or all anticlockwise: otherwise it will contain $\vec{r}, \vec{s}$ such that $\vec{r} \wedge \vec{s}$ is one of the six segments, putting $\{\vec{r}, \vec{s}\} \subseteq \tau$ in $\mathcal{F}_n = \mathcal{F}$ with a contradiction. But if τ orients all diagonals clockwise, say, then $\bigcap \tau$ is one of the segments, so $\tau \in \mathcal{F}$ since $|\tau| = 3$, again with a contradiction. So S has no $\mathcal{F}$-tangle.

   But neither is there an S-tree over $\mathcal{F}$. Indeed, as every element of $\vec{S}$, viewed as a subset of V, has more than n elements, our $\mathcal{F} = \mathcal{F}_n$ contains no singletons $\{\vec{s}\}$. But any S-tree over $\mathcal{F}$ needs singleton elements of $\mathcal{F}$ at its leaf nodes, those incident with only one edge $s \in S$.

   Since any tree set over $\mathcal{F}$ would define an S-tree over $\mathcal{F}$, see [8], this also shows that a feature system need not admit an $\mathcal{F}_n$-tangle or contain a tree set over $\mathcal{F}_n$; see two footnotes up.

15. Note that σ is a star as soon as every two of its elements form a star, which happens by construction. See [12, Lemma 25] for details, given in different notation.

Notes for Chapter 12

1. As long as p has answered T consistently with the people in V, there will be such a tangle of S. It will be unique, since T distinguishes all the tangles of S.

2. To make this example more extreme, add two people to V whose views are the exact opposite of this popular tangle τ, and assume that no two of the other people in V agree on all of S. (This can well happen even though each of them agrees with τ on 90% of S: all we need is that the number of 10% subsets of S is bigger than $|V|$.) Then this opposite view would be chosen as σ, being the only specification of S supported by two or more $v \in V$, although its guess is 90% wrong for all but two of the people in V.

3. The specification $p(S)$ will be consistent, by definition, unless p has some three features not shared by any $v \in V$. If this happens, chances are that either p has some rare medical condition that differs from the closest common illness by one of these three features, or that our measurement for one of those three was erroneous.

4. Recall that we excluded the possibility that a tangle answered 'no' to all four questions at a given position, by placing the corresponding set of four no-answers in $\mathcal{F}$.

5. For example, when we seek to compare viruses that mutate fast, alignment may be impossible or unreliable even when they originate from a common recent ancestor.

6. More formally, one can prove that there exists a similarity measure σ on $V^{(2)}$ such that for the set S of all partitions of V the tree of all the tangles in $\vec{S}$ with respect to the order function (O1) based on σ is exactly the Darwinian tree of life for the organisms or species in V. This similarity measure σ can, in principle, be computed without prior knowledge of the Darwinian tree of life, which is thus determined by this tree of tangles.

7. Does 'similar' mean 'pairwise similar'? Or should a group in W be the transitive closure of the set of pairs of similar elements? Or neither?

8. If we view every feature $\vec{s} \in \vec{\tau}^{+}$ as the set $\{ v \in V \mid v(s) = \vec{s} \}$, as usual, then $\bigcup \vec{\tau}^{+}$ is the union of all those sets.

9. Notice the analogy with Wittgenstein's problem in capturing the extensionality of meaning, as described in Section 5.3. Although the set of things a word names is given, somehow, in terms of the sets of things named by certain predicates, the exact relationship between the two is neither that of just union or intersection of sets.

10. It need not guide the specification, as $\overleftarrow{v}$, of those v that do not lie in $V(D_\tau)$: if $v \in V$ has fewer than $p \, |\tau^{+}|$ of the features in $\vec{\tau}^{+}$, it need not lack at least $p \, |\tau^{+}|$ of those features. But we do not care about those v; what matters are those in $V(D_\tau)$.

11. This will usually happen, and always if $|F|$ is odd, if 'most' just means 'more than half'. If 'most' means 'at least a proportion of p' with $p > \frac{1}{2}$, it need not hold.

Notes for Chapter 13

1. Locations of tree sets were defined in Section 8.1.

2. However, remember that the questions in T are not necessarily in our original questionnaire S: they may be combinations of such questions. But in order to distinguish all the mindsets on S, it is enough to ask the questions from S needed to build the composite questions in T, which may be far fewer than all of S.

3. In the graph-theoretical tree $\mathcal{T}$ representing T as in Section 8.1, the above sum is over all the edges of $\mathcal{T}$ *not* on the path between the two nodes that correspond to the maximal tangles in $\vec{S}$ that extend $p(T)$ and $q(T)$, respectively.

4. It is important here not to confuse concepts with words used to express them: since language is defined by interpersonal consensus, the individual's concepts and notions we are trying to understand here will typically not be associated with a word.

5. Compare the discussion of this point in Section 12.2.

6. Older students might, of course, simply be interviewed and asked which teaching methods they prefer. This may be less significant, but it may help make such a study feasible.

Notes for Chapter 14

1. formally: contain as an element

2. We proved this in Section 2.4, and again in Section 7.3.

3. For example, an element $v \in V$ is more similar in this ordering to a 6-tangle ρ than to a 9-tangle τ if ρ and τ agree with v on equally many s of orders 0, 1 and 2, respectively, but more s of order 3 satisfy $\rho(s) = v(s)$ than $\tau(s) = v(s)$. It is then irrelevant how ρ, τ or v specify any s of order > 3.

4. Formally, we extend our tree set T to a tree set T' of partitions of V in such a way that no $v \in V$ lives at a location of T' that is not also a location of T. This can be done step by step, as in the algorithmic proof in Section 7.3 of why the principal orientations τ_v of the set of all partitions of V are its only tangles.

5. To see this, imitate the proof in either Section 2.4 or 7.3 of why the principal orientations τ_v of the set of all partitions of V are its only tangles.

6. The arbitrariness of this grouping also shows up in the fact that there are many ways of choosing those additional blue partitions, which would divide the seven real clusters into different groups not grounded in reality.

7. See Note 11 in Section 2.4 and Figure 7.4 for why large S ensure this.

8. We might ask in addition that X should witness τ, perhaps even minimally so, subject to maintaining its reliability as a guide for τ, but our examples so far do not seem to require or suggest this.

9. We follow the custom amongst computer scientists of drawing such trees upside down, to illustrate the decision process that follows the questions $q_1, \ldots, q_m$ in steps.

10. The *decision* tree for the set $T \subseteq Q$ of questions is very different from the graph-theoretical tree that captures the structure of T as in Figure 8.1. That tree is not binary, and the tangles of T can live in any of its nodes, not just in its leaves.

11. The black hole tangle was a simple example.

12. In fact, f_1 was used for reference only, such as when we required the other f_i to be orthogonal not only to each other but also to f_1. Unlike $f_2, \ldots, f_\ell$, it was not chosen as a feature for $\vec{S}$. But there is no harm in treating f_1 like all the other f_i here, so for simplicity we shall do so.

13. This is true even if we extended $\vec{S} = \{f_1, \ldots, f_\ell\}$ not only by adding inverses, as we must in order to obtain a feature system, but also some infima or suprema. That is because how a tangle specifies $\vec{r} \wedge \vec{s}$ and $\vec{r} \vee \vec{s}$ is determined by how it specifies $\vec{r}$ and $\vec{s}$: only one specification of $\vec{r} \wedge \vec{s}$ and of $\vec{r} \vee \vec{s}$ will be consistent with its specifications of $\vec{r}$ and $\vec{s}$.

14. Being distinct tangles of the same order, τ' and τ'' are also distinguishable, even though neither is distinguishable from τ. Thus, being indistinguishable is not an equivalence relation, neither for tangles nor for regions (see below).

15. Since our approach is generic, just in terms of tangles, it can be applied directly to generic clusters too. This, then, gives rise to notions of complexity, cohesion and visibility for generic clusters exactly as outlined below for regions. Since these notions have their origin in image analysis by tangles [16], however, and 'visibility' has its most natural examples in this context, we treat them here.

16. Distinct regions containing tangles of the same order are distinguishable, of course, because distinct tangles of the same order are distinguishable.

17. Our notion of 'equivalence' for tangles is indeed an equivalence relation on the set of tangles in $\vec{S}$ in mathematical terms. The proof is pretty and not entirely trivial. One aspect of mathematical equivalence, called transitivity, implies that if ρ_1 is equivalent to ρ_2 and this is, in turn, equivalent to ρ_3, then ρ_1 *must* be equivalent to ρ_3 too.

18. This means, mathematically, that it is invariant under the automorphisms of the feature system $\vec{S}$. It implies that an algorithm that computes this tree of tangles T returns the same T no matter how we present S to it – for example, in which order we feed the partitions $s \in S$ to our computer, or whether we present s as the set $\vec{s}$ or as its complement $\overleftarrow{s}$.

19. Note that, unlike in most other contexts, every $s \in S$ has a default specification here: we use $\vec{s}$ to indicate that s is 'present' rather than 'absent' in the purchase.

20. It is not quite as bad as this. If ten percent of the items in S are 'green', and ecologically-minded customers buy only green items, then the above calculations yield that an average triple of green items is bought by $1/1000$ ecologically-minded customers, as well as a few others, rather than by one in a million. So we may have a 'green' tangle as soon as n is about $|V|/1000$, and we may only need a few thousand customers for this to work.

21. In our case, these are the tangles $\{\,\bar{s}\mid s\in S\,\}$ of S and $\{\,\bar{v}\mid v\in V\,\}$ of V.

22. ... albeit, in our example, of the dual feature system. One can equally well consider tangles of partitions of V with the biased version of (O1) as the order function.

23. Such algorithms have existed for some time and continue to be developed and improved; they are not the topic of this book.

24. Imagine a party game where a person thinks of something and we have to guess what it is by asking yes/no questions of any kind. Every such question is a 'potential feature'.

25. See Section 6.5 for a discussion of the quality of notions.

26. Compare the beginning of Section 9.3.

27. This is possible: some words have several meanings so different from each other that the word fields in which they appear are disjoint except for this word.

References

[1] Tangles: e-book and software. Home page at `tangles-book.com`.

[2] H.v. Bergen. Abstract tangle-tree duality. In preparation, 2023.

[3] H.v. Bergen and R. Diestel. Traits and tangles: an analysis of the Big Five paradigm by tangle-based clustering. In preparation.

[4] J. Carmesin, R. Diestel, M. Hamann, and F. Hundertmark. Canonical tree-decompositions of finite graphs II. Essential parts. *J. Comb. Theory Ser. B*, 118:268–283, 2016.

[5] J. Carmesin, R. Diestel, M. Hamann, and F. Hundertmark. Canonical tree-decompositions of finite graphs I. Existence and algorithms. *J. Comb. Theory Ser. B*, 116:1–24, 2016; arXiv:1406.3797.

[6] J. Carmesin, R. Diestel, F. Hundertmark, and M. Stein. Connectivity and tree structure in finite graphs. *Combinatorica*, 34(1):1–35, 2014; arXiv:1105.1611.

[7] R. Diestel. *Graph Theory* (5th edition). Springer-Verlag, 2017. Electronic edition available at `http://diestel-graph-theory.com/`.

[8] R. Diestel. Tree sets. *Order*, 35:171–192, 2018.

[9] R. Diestel. Abstract separation systems. *Order*, 35:157–170, 2018; arXiv:1406.3797.

[10] R. Diestel, C. Elbracht, and R.W. Jacobs. Point sets and functions inducing tangles of set separations, 2021. arXiv:2107.01087; to appear in Journal of Combinatorics.

[11] R. Diestel, J. Erde, C. Elbracht, and M. Teegen. Duality and tangles of set partitions. *J. Combinatorics*, 15(1):1–39, 2024; arXiv:2109.08398.

[12] R. Diestel, J. Erde, and D. Weißauer. Structural submodularity and tangles in abstract separation systems. *J. Comb. Theory Ser. A*, 167C:155–180, 2019; arXiv:1805.01439.

[13] R. Diestel, F. Hundertmark, and S. Lemanczyk. Profiles of separations: in graphs, matroids, and beyond. *Combinatorica*, 39(1):37–75, 2019; arXiv:1110.6207.

[14] R. Diestel and S. Oum. Tangle-tree duality in graphs, matroids and beyond. *Combinatorica*, 39:879–910, 2019. arXiv:1701.02651.

[15] R. Diestel and S. Oum. Tangle-tree duality in abstract separation systems. *Advances in Mathematics*, 377:107470, 2021; arXiv:1701.02509.

[16] R. Diestel and G. Whittle. Tangles and the Mona Lisa. arXiv:1603.06652.

[17] C. Elbracht and J. Kneip. A canonical tree-of-tangles theorem for structurally submodular separation systems. *Combinatorial Theory*, 1(5), 2021.

[18] C. Elbracht, J. Kneip, and M. Teegen. A note on generic tangle algorithms. arXiv:2005.14648.

[19] C. Elbracht, J. Kneip, and M. Teegen. Tangles are decided by weighted vertex sets. *Advances in Comb.*, 2020:9, 2020; arXiv:1811.06821.

[20] C. Elbracht, J. Kneip, and M. Teegen. Trees of tangles in abstract separation systems. *J. Comb. Theory Ser. A*, 180:105425, 2021; arXiv:1909.09030.

[21] C. Elbracht, J. Kneip, and M. Teegen. Obtaining trees of tangles from tangle-tree duality. *J. Combinatorics*, 13:251–287, 2022; arXiv:2011.09758.

[22] Bernhard Ganter and Rudolf Wille. *Formal Concept Analysis*. Springer, 1999.

[23] C. Godsil and G. Royle. *Algebraic Graph Theory*. Springer-Verlag, 2001.

[24] Martin Grohe and Pascal Schweitzer. Computing with tangles. In *Proceedings of the Forty-seventh Annual ACM Symposium on Theory of Computing*, STOC '15, pages 683–692, New York, NY, USA, 2015. ACM.

[25] M. Hermann and H.v. Bergen. Matrix-based order functions for set partitions and functions. In preparation.

[26] F. Hundertmark. Profiles. An algebraic approach to combinatorial connectivity. arXiv:1110.6207, 2011.

[27] Alexander K. Kelmans and Boris N. Kimelfeld. Multiplicative submodularity of a matrix's principal minor as a function of the set of its rows and some combinatorial applications. *Discrete Math.*, 44(1):113–116, 1983.

[28] S. Klepper, C. Elbracht, D. Fioravanti, J. Kneip, L. Rendsburg, M. Teegen, and U.v. Luxburg. Clustering with tangles: Algorithmic framework and theoretical guarantees. *Journal of Machine Learning Research*, 24:1–56, 2023; arXiv:2006.14444.

[29] U.v. Luxburg. A tutorial on spectral clustering. *Statistics and Computing,* 17(4), 2007. arXiv:0711.0189.

[30] Sang-il Oum and Paul Seymour. Approximating clique-width and branch-width. *J. Comb. Theory Ser. B,* 96(4):514–528, 2006.

[31] N. Robertson and P.D. Seymour. Graph minors I–XX. *J. Comb. Theory Ser. B,* 1983–2004.

[32] N. Robertson and P.D. Seymour. Graph minors. X. Obstructions to tree-decomposition. *J. Comb. Theory Ser. B,* 52:153–190, 1991.

[33] J. Shi and J. Malik. Normalized cuts and image segmentation. *IEEE Transactions on Pattern Analysis and Machine Intelligence,* 22(8):888–905, 2000.

[34] Y.C. Wei and C.K. Cheng. Towards efficient hierarchical designs by ratio cut partitioning. *Proc. IEEE Int. Conf. on Computer-Aided Design,* pages 298–301, 1989.

[35] Ludwig Wittgenstein. *Philosophical Investigations.* Basil Blackwell, Oxford, 1953.

[36] A. Zielezinski, S. Vinga, J. Almeida, and W.M. Karlowski. Alignment-free sequence comparison: benefits, applications, and tools. *Genome Biology,* 18:186, 2017.

Index

Page numbers in italics refer to definitions. The alphabetical order ignores letters that stand as variables; for example, 'k-tangle' is listed under 't'.

Symbol Index

The entries in this index are divided into two groups. The first group shows symbolic notation that may involve variables. These are not part of the symbol being defined and may be different in the text. For example, if τ is a tangle of a set of partitions, and s is such a partition, then $\tau(s)$ denotes the side of this partition that lies in this tangle. While the list below indicates this notation by listing $\tau(s)$, it is deemed to cover also $\sigma(r)$ when σ is a tangle orienting a partition r – but not $f(x)$ when f is a real-valued function and $x \in \mathbb{R}$.

The letters in the second group are fixed, such as L for the Laplacian.